METMEN IN WARTIME

METEOROLOGY IN CANADA 1939–1945

Clockwise from upper left:
John Patterson,
Andrew Thomson,
J. Reginald H. Noble,
Patrick D. McTaggart-Cowan
[COURTESY: METEOROLOGICAL SERVICE OF CANADA]

METMEN IN WARTIME

METEOROLOGY IN CANADA 1939–1945

MORLEY THOMAS

ECW PRESS

NATIONAL LIBRARY OF CANADA CATALOGUING IN PUBLICATION DATA

Thomas, Morley, K., 1918 –
Metmen in Wartime: meteorology in Canada 1939–1945

Includes bibliographical references and index.

ISBN 1-55022-448-4

1. Military meteorology – Canada – History – 20th century. 2. Canada. Meteorological Division – History. 3. World War, 1939–1945 – Canada I. Title.

QC857.C3T59 2001 355.3´432 C00-933131-X

Cover and text design by Tania Craan
Layout by Mary Bowness
Printed by AGMV

This book is typeset in Sabon and Penumbra

Distributed in Canada by General Distribution Services,
325 Humber Blvd., Toronto, Ontario M9W 7C3

Published by ECW PRESS
2120 Queen Street East, Suite 200,
Toronto, Ontario, M4E 1E2
ecwpress.com

The publication of *Metmen in Wartime* has been generously supported by The Canada Council, the Ontario Arts Council, and the Government of Canada through the Book Publishing Industry Development Program. Canadä

Assistance has also been provided by Environment Canada.

PRINTED IN CANADA

DEDICATION

This book is dedicated to the senior meteorologists who led the Meteorological Division during World War II. They administered the provision of meteorological services for the Royal Canadian Air Force in the defence of Canada, in the protection of north Atlantic shipping and in the British Commonwealth Air Training Plan, as well as for the Royal Air Force in ferrying aircraft to Britain. Public weather forecasting and services for civilian air transport were maintained during the war that saw the size of the Meteorological Division increase tenfold.

John Patterson (1872-1956), Controller (head) of the Meteorological Division.

Andrew Thomson (1893-1973), Deputy Controller of the Meteorological Division.

Patrick D. McTaggart-Cowan (1912-1997), Principal Meteorological Advisor to the Royal Air Force Ferry Command.

J. Reginald H. Noble, Meteorological Advisor to the Royal Canadian Air Force Chief of Air Staff.

TABLE OF CONTENTS

PREFACE

This book deals with meteorology in Canada during the 1939 to 1945 period of World War II. It follows Forecasts for Flying (1996), a book on the history of Canadian meteorology during the 1918 to 1939 period. The latter, and an earlier book, The Beginnings of Canadian Meteorology (1991), were also published by ECW PRESS with assistance from the Meteorological Service of Canada.

Several years after his retirement as Controller of the Meteorological Division, John Patterson completed a manuscript entitled "History of the Meteorological Service of Canada during World War II" [c1955], but the manuscript was never edited nor published. The Patterson manuscript has been of great value to me in researching for this book in the Archives and Library of Environment Canada at Downsview.

Publication has been made possible through the cooperation of Dr. M.D. Everell, Assistant Deputy Minister responsible to Environment Canada for the Meteorological Service of Canada, and his immediate predecessor, Dr. Gordon A. McBean. I have enjoyed accommodation in the Downsview library and assistance from Ms. Maria Latyszewskyj, librarian, and her staff. The maps were drawn by William Johnson. Editorial assistance from Tracey Millen, ECW editor, and Jack David, publisher, is gratefully acknowledged.

Morley Thomas
Toronto, 2001

CHAPTER 1

AUGUST 1939

In August 1939, John Patterson, the head of Canada's national meteorological service, must have had great pride in his staff for the services they were providing to commercial aviation and the general public across Canada. The Meteorological Service had survived the great economic Depression but it had not been easy since the budget had been cut by 25 percent over three years and salaries had been reduced by 10 percent. Because of this, an initial attempt in the early 1930s to provide aviation weather services had to be abandoned and the newly hired airport staff dismissed. But, despite the Depression, the need for aviation weather services had continued to increase.

In 1936, when the federal government created a new Air Services Branch in the Department of Transport, the Meteorological Service was moved to this branch, renamed the Meteorological Division, and given resources to hire additional staff. New airport observing stations and forecast offices were established to serve the Trans-Canada Airlines, which was being organized at that time. By August 1939, Patterson's young staff had inaugurated a daily, transcontinental, twenty-four-hours-a-day aviation weather service for the new national airline and other commercial aviation. Further, on the East Coast, his meteorological staff had inaugurated an aviation weather service for summertime British and American experimental transatlantic flying boat flights.

Nearly seventy years earlier, in 1871, Professor George Templeman Kingston of the Toronto Observatory had been granted funds to establish a national meteorological service and within a few years, had commenced issuing storm warnings and public weather forecasts. By 1903, the weather observing network extended from coast to coast and weather forecasts were issued daily for all regions of southern Canada. But forecasting methods, based on the personal experience of the forecasters and their ability to project current weather conditions into the future using empirical means, remained unchanged and there had been scarcely any improvement in the forecasts over the decades. In the early 1930s, the forecaster could do little more for the aviator than provide him with the ordinary public weather forecast. The ever-expanding aviation industry demanded better and more suitable forecasts.

Fortunately, new physical theories in meteorology had been developed roughly twenty years earlier by Scandinavian meteorologists, and this opened the door for improved weather forecasting. The new methods had not been adopted by the Canadian public weather forecasters but John Patterson made sure that the new theories and methods became the basis of the instruction given in a Master of Arts degree course in physics (meteorology) first offered at the University of Toronto in 1933–1934. Consequently, when financial resources became available within a year or two, there were graduates with a sound understanding of the new scientific meteorological theories available for employment. By August 1939, when the Meteorological Division employed a staff of fifty-nine meteorologists, thirty-four of these were graduates of this new program and nearly all were employed as aviation forecasters at five airports — Vancouver, Winnipeg, Toronto, Montreal, and Gander.

To support this aviation weather services program, a new network of observing stations had been established at the airline's terminal and intermediate airports, and at the new radio range stations along the transcontinental airway. Thirty-five stations in this network, established during the 1936 to 1939 period, reported complete weather observations every hour of every day by teletype. These observations were indispensable to the weather forecasters, the airline dispatchers, and the pilots flying the airways. Valuable as the observations were to the airline personnel, however, the weather forecasts were becoming of greater importance as the flights were increasing in length, duration, and altitude.

In the 1930s, "meteorologist" was a new term and profession in Canada. Over the decades the men preparing the public weather forecasts had been classified as "weather forecasters." In 1930, when the Meteorological Service first hired university graduates to train as aviation weather forecasters for the airmail weather service, the term "meteorologist" had been used and its use continued within the Service. However, at the airports, it was the pilots and others concerned with scheduled civilian flying who first began to call the meteorologist the "metman," although it was in World War II that the Royal Canadian Air Force aircrew and others used the term almost exclusively for the civilian meteorological officers. Accordingly, the term "metmen," which became such a part of wartime meteorology, is used in this book to refer to the hundreds of civilian meteorological personnel who served with the air force in Canada during the war. There were some MA graduates or equivalent meteorologists at RCAF operational stations but the meteorologists were largely stationed at the main civilian aviation forecast offices. The great majority of metmen with the RCAF during the war were BA graduates.

Evidence of John Patterson's pride in his staff and their work in the late 1930s was his invitation to the two premier meteorological societies in the world, the Royal Meteorological Society of Britain and the American Meteorological Society of the United States, to hold a joint meeting in Canada. The meeting, the first ever for either society to hold outside their home country, took place in Toronto late in August 1939. Several world-famous meteorologists attended and arrangements were made to allow many of Canada's young meteorologists to also attend. But, there were clouds on the horizon that August — not meteorological clouds, but war clouds — as the political situation in Europe deteriorated. Within days of the end of the Toronto conference, World War II commenced and this would be a war in which meteorologists and metmen were to play a significant role.

CHAPTER 2

PREWAR METEOROLOGY

By 1939, considerable progress had been made in applying new scientific theories in meteorology to the forecasting of weather. The requirements of commercial aviation for better and more meaningful meteorological services than those that had been available in the early 1930s had been a forcing factor in this development. Along with the provision of the traditional public weather forecasts, the national meteorological service had begun enlarging its observing system and issuing aviation weather forecasts in 1937 when transatlantic flying began. By 1938, when Trans-Canada Airlines began scheduled transcontinental services, aviation meteorology had become the major program in the Meteorological Division.

EARLY METEOROLOGY

Descriptive meteorology had existed for millennia but the science of meteorology had developed very little before 1800. Then, half a century later, with the invention of the telegraph, national meteorological services were established in several countries, observations were collected from networks of stations, and empirical methods of weather forecasting were developed. In the United States, Congress ordered the establishment of a national weather service in 1870 to produce storm warnings for the Great Lakes and the Eastern seaboard. The next year, in Canada, Professor G.T. Kingston, director of the Toronto Observatory, obtained funds from the Department of Marine and Fisheries to organize a climate observing program in the four Canadian provinces. But, the Americans needed observations from Canadian stations to enable them to prepare meaningful storm warnings and the Canadian government was eager to get warnings for Canadian ports and harbours. So to ensure continuing government support, Kingston was obliged to make his main effort the organization of a Canadian network of telegraphic reporting stations. In this way, a Canadian Meteorological Service came into being on July 1, 1871, and a daily exchange of observations with the Americans began in January 1872.

Not content to simply relay the Washington-made storm warnings to Canadian ports and harbours on the East Coast and along the Great Lakes-St. Lawrence waterway, Kingston requested additional resources to enable him to prepare the warnings in his small Toronto observatory. Within a few years Kingston was given an increased allocation and, with additional staff, he began to prepare experimental forecasts. By 1876, the staff had gained sufficient experience to commence issuing Canadian-made storm warnings and weather forecasts to the public for the populated regions of Ontario and Quebec, and this service was extended to Atlantic Canada by 1880.[1]

In the 1870s, the national meteorological directors in Europe organized an International Meteorological Organization (IMO) to facilitate formation and adoption of common definitions, methods of observation, and telegraphic weather codes. A free international exchange of current weather observations began and continues to the present. Canada was not an active participant in the IMO for the first twenty or so years but, along with the United States, accepted most of the IMO definitions, methods, and codes. This international amity encouraged the national weather services in Europe and North America to expand their meteorological observing networks and improve their techniques of synoptic weather analysis and forecast preparation over the succeeding decades.

PREWAR METEOROLOGY

PUBLIC WEATHER FORECASTING

Over the pre-Great War decades, various theories of storms were developed, analysis techniques were honed, and even statistical methods of weather forecasting were attempted. But, forecasting remained largely empirical, almost a pseudo-scientific art, and the forecasters met with limited success at best in providing the public with useful and timely forecasts. There were two chief reasons for this mediocre success.

First, although a network of observing stations had been put in place, there were no regular upper air observations that would allow a daily three-dimensional description and analysis of the atmosphere. Second, there existed no sound scientific theory by which the movement of large atmospheric pressure systems could be forecasted or even explained. Hence, forecasting experience and inspection of past weather maps to determine any relevance to the current situation remained the best forecasting tools available to a weather forecaster. While the empirical methods had been improved somewhat over the years, the original weather forecasting system developed in the 1870s and 1880s was still in use in the Toronto and Victoria public weather forecast offices in 1939.

Twice daily, in the morning and again in the evening, the observations from scores of stations in Canada and the United States were plotted on maps by assistants. The maps were studied by the forecaster who drew isobars (lines of equal sea-level pressure) and marked areas of precipitation while considering probable developments over the next day or so. Once he had decided how the weather systems would develop and move he prepared next-day forecasts for specific regions of the country. The Toronto forecast office, called the "map room," was located at the Central or Head Office in the Observatory at 315 Bloor Street West. The Victoria, British Columbia, forecast office was at the Gonzales Heights Observatory in that city. The Toronto office prepared and issued forecasts twice a day for several regions across southern Canada, from the Rocky Mountains to the Atlantic Ocean, while a much smaller forecast office at Victoria issued forecasts for the southern part of British Columbia.

The network of telegraphic reporting stations had been expanded a little in the 1930s when the service obtained extra resources for aviation meteorology. In 1939 there were ninety or so of these part-time stations where the observers took two or three observations a day and sent them to the Toronto Central Office. The stations in the South used local telegraph offices for transmitting their observations while the stations in the subarctic sent their observations by radio to specific points on the telegraph system for relay to the Toronto and Victoria public forecast offices.

The forecasters needed American as well as Canadian observations to analyze weather conditions over the continent, and since 1872 there had been a free and unhindered exchange of meteorological observations between the two national meteorological services. This traditional exchange originally consisted of the morning and evening synoptic observations. The observations were taken at 7:25 a.m., 4:25 p.m., and 11:25 p.m. Toronto time but the afternoon observations were not exchanged.

Until the mid-1930s, the Canadian observations were collected at the Toronto Observatory, checked for reasonability, and, if acceptable, were included in a collective message and sent to Buffalo for relay to Washington. In return, a collection of American observations was sent from Washington through Buffalo to Toronto. There was another exchange on the Pacific coast between Victoria and Seattle, Washington.

Some of the Canadian public weather forecasters were university science graduates but they were not familiar with, and never used, the new theories and methods being developed in Europe. In Canada these were first used in the new University of Toronto Master of Arts program in 1933–34. There is no evidence that the public weather forecasters were interested in learning of the new developments, although Bernhard Haurwitz, a European-trained physicist at Toronto for a few years, later wrote that one or two of them actually came to a lecture by Vilhelm Bjerknes, the Scandinavian scientist responsible for the new methods of analysis and forecasting.[2]

There is no record of John Patterson planning to staff the map room with meteorologists trained in the new methods. But, as it happened, the escalating demands for meteorologists from both civil and military aviation during wartime forced him to delay any plans he might have had in this regard.

PREWAR METEOROLOGY

BERGEN SCHOOL OF METEOROLOGY

During the Great War, the Scandinavian countries, denied daily weather information from Britain and other western European countries, received only observations from stations in their own area to use in weather forecasting. Also, normal trading patterns were disrupted and Norway found it difficult to import sufficient quantities of foodstuffs for domestic consumption. To help increase home food production, a physicist, Professor Vilhelm Bjerknes, was employed to seek a method for improving weather forecasts for fishermen and farmers in the country. A pioneer in theoretical physical hydrodynamics, Bjerknes undertook to apply the principles of his science to the atmosphere. He and his associates were successful in their research and developed methods and ideas that were labelled "Polar Front Meteorology." Using this theory, with observations from an improved network of closely located stations, he was soon able to make improved forecasts available to Norwegians.[3]

The Bergen School ideas and methods pertaining to the development and motion of weather systems were published after the war and were used in several European countries. But the Bergen work was largely ignored in the national weather services of Canada and the United States for many years. There was the belief that, since the western European air masses were usually moister than those in eastern North America, the new theories could not be applied to North American weather systems.

But, at least two Canadian meteorologists, Andrew Thomson and Donald Archibald, became interested enough in the Bergen developments to go, at their own expense, to Norway and Germany for extended periods in the early 1930s to learn more about the new methods. By that time, three American universities, the Massachusetts Institute of Technology, the California Technological Institute, and New York University, had begun to offer courses in the new meteorology. In Canada, the University of Toronto, prodded by John Patterson and Andrew Thomson, offered the first Canadian graduate course in physics (meteorology) in 1933–34 in which the Bergen School theories and methods were introduced and used.

AN AIR SERVICES BRANCH

During an earlier period, from 1928 to 1932, the Post Office contracted with aviation companies to fly the mail between major cities on the Prairies and in southern Ontario and Quebec. The Meteorological Service had been given resources to organize and provide what was then called an "airmail weather service" for the aviation contractors. At twelve airports along the airways, weather observers had been hired by the Meteorological Service to observe the weather and use a Department of National Defence teletype system to distribute the observations. Also, the Meteorological Service was authorized to hire scientists for training as meteorologists who could provide the pilots with aviation forecasts, forecasts much more to their needs than the available public weather forecasts. But, the economic Depression deepened, the airmail contracts were cancelled, and funding for the airmail weather service eliminated. All but one airport weather office (Montreal St. Hubert) were closed and most of the new airport weather observers were dismissed.

By the mid-1930s, despite the Depression that still gripped the country, commercial aviation again began to grow rapidly and with this expansion the need for meaningful aviation meteorological services became pronounced. Faced with the necessity of either establishing a Canadian national airline with attendant facilities and services, or of watching the nation's air transport needs handled increasingly by the northward expansion of United States airlines, the federal government began to take decisive action.

In 1936 a new federal Department of Transport was formed which included an Air Services Branch dedicated to provide facilities for commercial aviation. Into this branch were gathered three national agencies from other departments which contributed services to commercial aviation — civil aviation, radio, and meteorology. Accordingly, the Marine Department's Meteorological Service became the Meteorological Division of Air Services and was given resources to expand its weather observing system and to hire aviation weather forecasters. At the same time, the new Meteorological Division was relieved of responsibility for the national Time Service and for all other non-meteorological activities.[4]

The next year, Trans-Canada Airlines (TCA) was formed and the

Meteorological Division became responsible for providing full aviation meteorology services for its operations. To serve TCA, and to a much lesser extent other commercial airlines and itinerant aircraft, the Meteorological Division established Dominion Aviation Forecast Offices (DAFOS) at TCA's divisional headquarters and meteorological observing stations at all terminal and "intermediate" airports where scheduled landings took place. Also, the Meteorological Division became responsible for providing meteorological instruments to the radio range stations at other airports along the transcontinental airway so that the radio operators could take observations at those stations. Dedicated meteorological teletype circuits were developed to connect all observing stations and the DAFOS.

TORONTO MA IN METEOROLOGY

John Patterson, director of the Meteorological Service, had seen an earlier attempt to provide an airmail weather service prove to be woefully inadequate and he was determined that, when such services were again demanded, he would have a properly trained staff in place to provide them. Patterson was a well-respected physicist and this aided him in successfully lobbying the University of Toronto, and especially its somewhat conservative physics department, into instituting a postgraduate degree course in physics (meteorology) during the depths of the great economic Depression. With the assurance that the Meteorological Service physicists would give the meteorological courses and the practical laboratory sessions, the university launched the course in the 1933–34 academic year. To spark the interest of students, there was the tacit understanding that employment in the Division might well become available for graduates.

To be admitted to the one-year postgraduate course, a candidate had to hold an undergraduate degree of high standing in honour mathematics and physics or in engineering physics. Accepted by the university, the student had to be prepared to spend an academic year in intensive study since several courses in theoretical and applied meteorology and in other advanced mathematics and physics subjects were included in the more than a dozen lecture hours each week. Also, the students spent four afternoons a week in laboratory training in weather map analysis and practiced aviation weather forecasting using sets of historical weather maps and, later in the course, current maps.

During the first year or so of the course, there were lectures by such Scandinavian experts in the new meteorological theory as Jacob Bjerknes and Sverre Petterssen, and a German physicist, Bernhard Haurwitz, arrived at the university on a Carnegie Institute fellowship. The presence of these physicists hastened the acceptance of the Toronto program and the new methods of weather forecasting in Canada.[5]

As it happened, the government's response to the demands of commercial aviation was not developing as quickly as perhaps Patterson had anticipated. No resources were provided for aviation meteorology in 1934 but the first three graduates of the course that year were hired for other duties at the Toronto Head Office. There were no positions immediately available for the 1935 graduates either but, in 1936, when the Department of Transport Act was proclaimed and the Meteorological Service was administratively transferred to a new Air Services Branch, more resources became available. These were used to hire all available graduates and to establish District Aviation Forecast Offices to provide services for a national airline.

Scheduled flying by Trans-Canada Airlines of mail and passengers began in 1937 and, by the summer of 1939, the new Meteorological Division had hired thirty-four of the thirty-nine graduates of the MA course. All but three of the graduates were employed in aviation meteorology at the new District Aviation Forecast Offices.[6]

OBSERVATIONS

Fortunately, the Depression had not forced a reduction in the extent of the existing network of sixty-five or so telegraphic and wireless weather reporting stations. Full-time Meteorological Division employees manned a few of these, but most observations were taken by local individuals hired on a part-time basis to take two observations a day and telegraph the data to Toronto.

Beginning in the 1920s, several stations had been established in the western subarctic where weather observations were taken at posts of the Royal Canadian Corps of Signals and transmitted south. By 1930, a similar situation existed around Hudson Bay where observing duties were carried out by staff at federal Marine Radio Division stations. In addition to this regular, twice-a-day collection of Canadian data, the forecast offices at Toronto and Victoria received observations twice daily from selected lists of United States Weather Bureau stations.

By the mid-1930s, additional resources had been given to the Meteorological Service to allow a modest expansion in the network of synoptic observing stations. Then, in 1937–38, with the establishment of the transcontinental airway for Trans-Canada Airlines operations, a major increase in the number of observing stations took place. At the four airports (Vancouver, Winnipeg, Toronto, and Montreal) where District Aviation Forecast Offices (DAFOs) were being established and at other airports where TCA had scheduled landings (Edmonton, Calgary, Lethbridge, Regina, Kapuskasing, North Bay, and Ottawa), the Meteorological Division installed its own weather observers. At the emergency landing strips, located about every 125 miles (200 km) along the airways, radio range facilities were installed by the department and the radio operators became responsible for taking weather observations.

Throughout Canada, by the summer of 1939, there were more than 135 operational weather stations monitoring and reporting the weather seven days a week. Ninety or so of these stations were the old type synoptic observing stations where traditionally two observations were taken each day. By 1939, however, many of these stations had been asked to increase their daily observing programs to three or four synoptic observations. At the thirty-one observing stations at airports and along the airway, both synoptic and hourly observations were taken.

The airport observers took "synoptic" or complete, detailed weather observations at four fixed times each day (0130, 0730, 1330, and 1930 hours). In addition, the observers took aviation weather observations every hour (sequences). Both the synoptic and sequence observations covered most of the same meteorological elements — temperature, humidity, pressure, wind, clouds, and visibility — but the manner in which the information was entered on the teletype was markedly different. In 1939, the synoptic code had become numeric and consisted of four or more groups of five digits each. (Previous to

July 1939, Canada and the United States had used a word code for synoptic messages although European countries had long used an internationally approved numeric code.) The sequence code was a combination of letters, digits, and symbols organized in a simple fashion so the uninitiated could quickly understand and use it.

AIRWAY COMMUNICATIONS

Since 1872 there had been a free and unhindered exchange of meteorological observations, taken at 7:25 a.m. and 11:25 p.m. Toronto time between Canada and the United States by telegraph. Over time, the hours of observation were altered somewhat and pilot balloon observations were added to the exchange. Then, after the introduction of teletype in the early 1930s and the beginning of radiosonde observations and of hourly observations along the airways in the immediate pre-war years, the exchange had grown rapidly. In addition to the Toronto-Buffalo link another exchange had been opened between Seattle and Vancouver. A few days after the war began, the Montreal St. Hubert DAFO was connected to Albany, New York, to further facilitate the exchange of observations and other meteorological information between the two countries.

Immediate communications were vital to the operation of the meteorological system and traffic on the teletype circuits was scheduled and rigidly controlled. At each DAFO there were four teletype operators to man three shifts a day. Besides handling the DAFO meteorological traffic and relaying observations and forecasts from one circuit to another, the operators monitored and managed around the clock all traffic inserted by the observing stations. Along the airways, the dedicated national teletype system was divided geographically into sections or sub-circuits, which were linked at the DAFO relay points. In the subarctic, the synoptic observations were radioed to a specified receiving station where the information was put on the teletype circuit.

The collection of observations began at specified times and the observing stations reported in prescribed order. Other time periods on the circuit were designated for the relay of observations from one region to another and for the transmission and relay of forecasts. In addition to the meteorological traffic, short periods were designated for administrative messages and for "flight plans" which were sent from one airport to another advising of the departure and arrival times of itinerant aircraft.

The hourly observations, usually not plotted on maps, were of value to the meteorologists since inspection of them allowed a fine-tuning of the forecast just before it was released and for verifying the forecast in the hours after issue. However, the basic tool of the forecaster was the synoptic map and a new one was plotted for analysis every six hours. At each District Aviation Forecast Office, some assistants were responsible for the plotting work and general assistance to the forecasters while others took the observations.

A continual watch on the weather was maintained and if a sudden change occurred or if the conditions deteriorated below prescribed levels, "special" observations were taken and put on the teletype circuit. The teletype operators were responsible for transmitting the local observations and the forecasts prepared by the meteorologists, for relaying information from one circuit to another, and for monitoring and maintaining the scheduled flow of traffic on each of the circuits.

In the summer of 1939, to maintain an around-the-clock, seven-days-a-week operation, each DAFO had an establishment calling for seven meteorological assistants to monitor the weather, take the observations, and plot the maps.

FORECASTS FOR TCA

As valuable as the weather observations were to the Trans-Canada Airline pilots and dispatchers in keeping them up-to-date on the weather at airports and other stations within a radius of a few hundred kilometres of their base, weather forecasts had become of even greater importance in airline operations. It was for this reason that District Aviation Forecast Offices had been established at flight terminals where pilots and dispatchers were briefed before each flight on current and forecast conditions over their scheduled routes.

By about an hour after each synoptic observation time the technicians had received, decoded, and plotted the data on a weather map. There was an internationally accepted format for plotting the synoptic observations around each station circle and, to speed recognition by the analyst of the plotted numbers and symbols, blue ink was used for some elements and red for others. The meteorological assistants became very proficient in mentally translating the data from the numeral code to the equivalent map figures and symbols and the plotting was done rapidly. The maps used in eastern Canada covered all of Canada and the United States while those used in the West extended from the Pacific Ocean eastward over the continent to about the 75th meridian.

As the meteorologist studied the data plotted around the station circles on the map, he drew isobars and marked the positions of centres of high and low pressure. He then drew in the positions of the fronts — cold fronts in blue and warm fronts in red (a front is defined as the interface or transition zone between two air masses of different density).[7] To allow rapid visual identification of regions with poor flying weather, areas of precipitation were usually shaded in green and areas of fog in yellow.

After the meteorologist-forecaster had inspected and analyzed the current map, he then prepared "progs" (prognostic charts) showing what he thought the weather maps would look like twelve and twenty-four hours ahead based on the motion and development he believed would likely take place. In this analysis the meteorologist made use of all the observed and derived weather information plotted around each station circle — temperatures; cloud types, amounts and heights; wind speed and direction; visibility; humidity; and pressure tendency. With his background of theoretical meteorology he studied these parameters which enabled him to prepare much better progs than would have been possible by empirically projecting the current conditions into the future.

Upper air observations enabled the forecaster to make full use of the new air mass and frontal theories in analysis and forecasting. Access to profiles of temperature and humidity at several stations plotted by assistants allowed the forecaster to make a three-dimensional analysis and provided important information on the stability of the air mass near each upper air station. There were no operational Canadian radiosonde stations in 1939 but the first such stations were being installed in the United States and the observed data from pertinent stations were transmitted to the Canadian teletype circuits.

The radiosonde was a balloon-borne instrument that simultaneously measured pressure, temperature, and humidity as the balloon ascended and transmitted radio signals representing the data to a ground station. At each upper air observing station these signals were translated into values of the meteorological elements by technicians and put on the teletype circuit for transmission to the DAFOs. At the forecast offices the data were plotted on tephigrams (thermodynamic diagrams indicating atmospheric stability) for use by the forecaster. In general, two observations a day were taken at each upper air observing station. At this time, there was not yet a large enough number of radiosonde station reports for the forecaster to attempt the construction of upper air maps.

With the current map, progs, tephigrams, and antecedent maps (for history), the forecaster completed his analysis and then wrote or typed out area and terminal forecasts for use in local briefing sessions and for transmission by the teletype operator to other offices and users. The area forecasts gave the general weather to be expected over the next 24 hours: the speed and direction of winds at the surface and at three or four levels above mean sea level; the cloud types, heights and amounts; the cloud ceiling; the visibility; and any expected obstructions to visibility.

A "remarks" portion of the forecast contained information about the movement of the major pressure systems and whether or not any fronts might affect the weather of the area. Terminal forecasts, issued for all airports where Trans-Canada Airlines had scheduled landings, were site specific and spelled out the expected weather at each airport in greater detail than that given in the area forecasts. These forecasts were made and issued every six hours every day, seven days a week.

It was mandatory, during the early days of TCA and the DAFOs, that the airline dispatcher and the pilots visit the meteorological office and receive a personal briefing from the meteorologist before every flight. Besides having the current weather and forecasts at the intended and alternate terminals explained, the crew was given a written forecast and a "cross-section" map depicting the weather vertically and horizontally, along the proposed route. During the flight, the pilot maintained contact with the radio range operators below his flight path and, as well as receiving reports from them, advised the range operators of any variation in the weather from that forecasted. The pilot made notes on a prescribed form and these "pireps" (pilot reports) were taken to the weather office at the end of each flight when the pilot would brief the meteorologist of the weather conditions he had encountered.

The first DAFO to open was the one at the Vancouver Municipal Airport on Sea Island where Allen McCauley began providing services for TCA's first scheduled flights to Seattle in September 1937. By March 1938, when TCA began to fly the main route between Vancouver and Winnipeg, a second DAFO was opened at Winnipeg's Stevenson Aerodrome (Field). The officer-in-charge at Winnipeg was Don Archibald who had been superintendent of the Western Air Mail Weather Service earlier in the decade and had returned to Winnipeg to co-ordinate the establishment of weather services for TCA. In October 1938, the main route was extended to Montreal and Toronto and a spur route was flown from Lethbridge on the main line to Calgary and Edmonton. A full-service DAFO was opened at Montreal St. Hubert airport that year under Fred Mahaffy and limited meteorological services were provided at Toronto Malton Airport. (Early in the fall of 1939 a full-time staff was posted to Malton with Fred Turnbull as officer-in-charge.) The transcontinental TCA route was extended to Moncton, New Brunswick, in September with forecasts prepared at Montreal since the Moncton DAFO would not be opened until meteorologists became available in May 1942.

In 1939, to monitor the weather, analyze a map and prepare route (or regional) and terminal forecasts every six hours each day, and to brief the pilot of each departing aircraft, there were but five meteorologists on each District Aviation Forecast Office staff.

DAFO STAFF

Such was the demand for aviation forecasters that just about all the thirty-nine prewar graduates of the University of Toronto MA degree course in meteorology accepted positions in the Meteorological

Division. When a new meteorologist was hired he sometimes remained in Toronto for a few weeks for more practical training or was sent to instruct new observers, especially radio range operators. On occasion, a graduate was kept in Toronto to assist in instructing students in the following course in weather map analysis. More often, the new employee was posted immediately to one of the new DAFOs where for several weeks he worked on shift with an experienced forecaster. Then he was given responsibility for his own shift at the same office or he was moved to a different one where a forecaster was urgently needed. The DAFOs were expanding and it was often necessary to move an individual from one office to another on short notice.

The meteorological assistants and teletype operators were not usually subject to so many different postings. Candidates for meteorological assistant positions, in which those accepted would become observers and/or map plotters, required high school graduation and were hired locally, as they were needed, and trained on the job by experienced staff at one of the DAFOs. (With the scarcity of jobs in the Depression several men hired as assistants were highly qualified university graduates and subsequently became meteorologists.)

Prospective teletype operators had to have typing skills and were hired locally and trained by experienced operators at the DAFOs. All positions were filled under Civil Service Commission procedures which meant that the positions had to be advertized for a few weeks, rating boards held, the names of successful candidates announced, and an appeal period allowed before anyone was actually hired. This process took weeks and sometimes months causing an annoying delay in establishing the DAFOs and getting them staffed for twenty-four-hour, seven-days-a-week operation.

TRANSATLANTIC FORECASTS

In Newfoundland, the Meteorological Division began providing an aviation meteorology service for transatlantic flying boat flights in 1937. At a November 1935 conference in Ottawa to consider creating an imperial transatlantic air service, agreement had been reached with the United Kingdom, Newfoundland, and the Irish Free State (Eire) that Canada would be responsible for the provision of weather forecasts for Newfoundland and the western half of the Atlantic. It was planned that a British company would undertake experimental summertime flights with flying boats and that an American company would be granted rights to make similar flights.

To lead the Canadian meteorological work, Patrick D. McTaggart-Cowan, then a Canadian Rhodes Scholar at the University of Oxford, was hired in 1936 to study with a small group of British Meteorological Office men, with similar future responsibilities, in order to develop methods and systems to best serve the experimental flights. In the early summer of 1937, McTaggart-Cowan set up a forecast office at the Botwood seaplane base in Newfoundland and the first flights by Imperial Airways and Pan American Airways took place in July of that year. There was but one flight in 1938 but the office remained open and daily weather maps were analyzed.

Since the governments planned for land-based aircraft to be eventually used for the transatlantic flights, construction had already begun on an airport near Gander Lake and, in December 1938, the meteorological office was moved to the newly completed Newfoundland Airport (later named Gander airport). In the summer of 1939, both Pan American and Imperial Airways operated scheduled flying-boat services from New York through Newfoundland and Shannon, Ireland, to Britain and return. (A government aviation reorganization in Britain at about that time had seen the name Imperial Airways changed to British Overseas Airways Corporation.)

Just before the declaration of war, early in September 1939, the transcontinental teletype network was extended to Moncton and Shediac, New Brunswick. The connection to Moncton was in anticipation of the extension of the transcontinental airway and TCA flights to the Maritimes, and an observing program was being organized at the radio range stations on the route east from Montreal. Shediac was included on the teletype circuit because a meteorological observer was located at that flying boat base to serve the Pan American flights on the route between Botwood, Newfoundland, and Port Washington on Long Island, New York.

The conditions under which forecasting for transatlantic flying were developed were quite different from those encountered in developing the transcontinental system for Trans-Canada Airlines.[8] While the transcontinental forecasters had weather reports available from airport stations spaced at 100-mile or so intervals (160 kilometres) along airways, there were no such regular weather reports from the Atlantic Ocean between Newfoundland and Ireland. The only observations from this vast ocean area were reports from a few itinerant ships and in-flight weather reports radioed to Gander from flying boat captains. Hence, the analysis of the existing situation and the forecasting of future weather over the Atlantic Ocean provided a remarkable challenge for the young meteorologists. The use of the new Polar Front meteorological theories and methods became invaluable.

Led by McTaggart-Cowan, the Canadian meteorologists were meeting this challenge in 1939. Carl Christie has noted in *Ocean Bridge* "By the time the war broke out in September 1939, McTaggart-Cowan was widely recognized as one of the foremost authorities on North Atlantic weather."[9]

ROYAL CANADIAN AIR FORCE

The Meteorological Service had only a few contacts with military aviation interests during the Great War and the situation changed little for some years. But, after the Royal Canadian Air Force was created in the mid-1920s, the force soon began to obtain additional resources for flying training and operations and more contacts were established with the Meteorological Service. Beginning in 1933, the RCAF annually sent two permanent force officers to Toronto for a week's familiarization visit at the meteorological headquarters. These contacts proved to be valuable to both organizations but, when the RCAF began to expand and requests for specific services were made, the Meteorological Service had neither mandate nor resources to respond.

When the newly named Meteorological Division became part of Transport's Air Services, however, the situation began to change. Early in 1937, the RCAF requested and obtained blank maps and codes to enable their technicians to utilize the international broadcasts of meteorological observations from Arlington, Virginia. Later that year, the Commanding Officer of a squadron based at Dartmouth, Nova Scotia, asked if the Meteorological Division could provide a course in meteorology similar to one given to a Winnipeg-based squadron in

1935. It was explained that this could not be done since the Service did not have a meteorologist at Halifax as it did at Winnipeg.[10]

After having weather maps plotted by technicians, the RCAF officers soon recognized that they did not have the personnel to properly analyze the data and make the forecasts they required. So, in the winter of 1938–39, the Department of National Defence made an official request to the Department of Transport for weather forecasts for the RCAF air bases and for areas 500 miles (800 kilometres) seaward from the coasts.[11] It was explained that this need would exist only until the RCAF could train its own forecasters. To aid in Transport's response to National Defence, John Patterson sent a copy of the regulations used in Britain where the civilian Meteorological Office provided all professional meteorological services to the Royal Air Force. Patterson recommended that this concept be adopted in Canada and his recommendation was subsequently accepted by Transport, National Defence, and the RCAF.

Consequently, in March 1939, as the RCAF accelerated the expansion of its operational activities in anticipation of possible hostilities, a request was made for a meteorologist and a technician for posting to the Western Air Command headquarters. Because of the lack of trained staff, Patterson had to advise the RCAF that this would not be possible before late summer. Then, more requests followed for meteorological personnel to be stationed as soon as possible at the air bases at Trenton, Ontario, and Dartmouth, Nova Scotia. Finally, in late August, the first civilian secondment of a meteorologist and an assistant was about to be made to the Western Air Command at Jericho Beach, Vancouver, British Columbia, but by then there were greater needs for the men and the secondment was cancelled.

In the summer of 1939, the RCAF's Air Training Command began to express an interest in meteorological instruction. On request, the Meteorological Division provided a list of the equipment previously sent to various stations. Patterson also sent a copy of *Instructions for Meteorological Reports on Airways* and urged the RCAF to use the airways instructions and the standard airways observing form rather than attempt to develop their own. He suggested that the Meteorological Division might train a few key observers at Toronto who would then act as instructors at the air bases, or that the Meteorological Division might take over all the meteorological work at the training bases as was being planned for the Jericho Beach operational station.

At that time, however, Patterson was using every available meteorologist to staff the new District Aviation Forecast Offices in order to provide round-the-clock service for Trans-Canada Airlines. And, since the RCAF was not yet sure of just what was needed, both organizations agreed that the plan for civilian participation in providing a meteorological service for the training command of the RCAF should be only slowly implemented.

METEOROLOGICAL DIVISION ORGANIZATION

The small Toronto Head Office staff, under the leadership of the Controller, John Patterson, was organized into sections with a meteorologist responsible for the various activities of each. In 1939, the Forecast Section, led by Frank O'Donnell, prepared the public forecasts for that part of the country from the Rocky Mountains to the East Coast; the Climatology Section, headed by A.J. Connor, administered the climatological station network throughout the country and collected, processed, and published climate data; and the Physics Section, headed by Andrew Thomson, was responsible for research, for aviation forecasting, and for lecturing and giving general assistance to the MA degree course. Smaller units looked after instruments, personnel, finance, stores, and the observations at the Toronto Observatory, with all heads reporting directly to Patterson. In 1939, the number of personnel at Head Office totalled slightly fewer than fifty people.

It is significant that there was no headquarters unit directly responsible for activities outside Toronto. From the beginning of the Meteorological Service in the 1870s, each part-time telegraphic reporting station observer and the officer-in-charge at each "outside office" (those stations with permanent full-time employees) reported directly to the director of the Service. The administering of these offices was of course done by several people in the Toronto Head Office sections but it was all done in the name of the Controller of the Meteorological Division. Patterson became the head in 1929 and signed all correspondence, believing that it was good managerial practice for each field officer to know that "John Patterson" was directly in control. Also, Patterson gained the satisfaction that he knew everything that was going on in the Meteorological Division. Further, it was rare that anyone but Patterson dealt with the department in Ottawa or with the International Meteorological Organization and foreign national meteorological services.

In retrospect, considering managerial theory, it is most unusual that no change was made in this singular and direct reporting arrangement when the transatlantic forecast centre began operating in Newfoundland in 1937 and when the District Aviation Forecast Offices opened in the next few years. Officers-in-charge were appointed by Patterson and, in most instances, each was the same man who had opened the office. In the rush of getting the offices open and the aviation forecasting system operating, no competitions were held nor was there any increase at first in classification or pay for the officers-in-charge. Most were expected to take normal forecasting shifts although as staffs grew in size, administrative duties expanded and regular shift work became impossible for an officer-in-charge.

This simple and direct management structure was in tune with other "Spartan" working conditions in 1939. Shift work was mandatory for the meteorologists and others at the aviation forecast offices, and the workweek consisted of five-and-a-half days for those at Toronto and at the "non-operating" outside offices. Pay amounted to less than $2,000 a year for all but a very few meteorologists and there was then no additional pay for shift work or overtime.

At the newly opened DAFOS, the twenty-four hours of the day had to be covered, usually with only four or five men in each category. When someone was sick or on annual leave, the others had virtually no time off as every eight-hour shift had to be covered each day and there was seldom a substitute meteorologist available. The young meteorologists and assistants accepted these working conditions since most had already suffered from the most severe economic Depression Canada and the world had ever experienced. Even for university graduates in such highly regarded specialities as mathematics and physics, jobs were most difficult to obtain. Every graduate in those days knew of several former classmates without jobs, and a position in a government service was highly valued.

Further, once a meteorologist had successfully completed the MA course and been hired and posted to a DAFO, he was usually inspired by the realization that he was participating in the application of a new scientific discipline and was challenged to help develop the new methods used in aviation forecasting. Few, if any, of the new meteorologists were married and thus each was able to accept a posting on short notice to a different part of the country.

Jobs were even more precious to the assistants than they were to the university graduates but usually the assistants were not asked to move around the country as frequently as the meteorologists were. These factors all led to a remarkably intelligent, well educated, and enthusiastic staff as the Meteorological Division moved into wartime.

A METEOROLOGICAL CONFERENCE

To assist in introducing the new meteorological theories and forecasting methods to Canadian meteorologists, Patterson had arranged for a few American and European scientists to visit Toronto in the 1930s and lecture to the Meteorological Office staff and to the MA students on course. Then, in 1939, when the Royal and American Meteorological Societies accepted John Patterson's invitation to hold a joint meeting in Toronto, it was to give the young Canadian scientists an opportunity to listen to and ask questions of the leading meteorological scientists in the world. Sixteen or so meteorologists from the United States, half a dozen from Britain and Europe, and two dozen Canadians met at the University of Toronto on August 28 and 29, 1939. The American group, led by Harvard Professor C.F. Brooks, included Commander Francis W. Reichelderfer, Weather Bureau chief, and W.M. Elsasser, H.R. Byers, V.K. Ekman, and others from the USWB. Scandinavians Jacob Bjerknes, Sverre Petterssen, and C.-G. Rossby were there while the British delegation included David Brunt, F.J.W. Whipple, and Sidney Chapman. With the war clouds thickening, J.M. Stagg, the leader of the British group, was prevented from attending since he was ordered to immediately return to Britain the moment his ship docked in Montreal.

After John Patterson's opening review of the history of Canadian meteorology, the formal sessions began and consisted of scientific, theoretical papers on radiation in the atmosphere, atmospheric radiation charts, extra-tropical cyclones, flow patterns in the atmosphere, characteristics of the general circulation, and horizontal divergence.[12] The papers were all theoretical and not one had to do with weather map analysis and aviation weather forecasting in which the young Canadian meteorologists were engaged. Since at this time there was very little research going on in the Meteorological Division or elsewhere in Canada, John Patterson apparently wished to have his young meteorologists exposed to the theoretical side of meteorology at the conference. This was characteristic of John Patterson; he believed that only excellent mathematics and physics graduates should be accepted as candidates for the MA course. Then, after indoctrination in the laws of physics as these apply to the atmosphere, the new meteorologists would be able to handle the practical aspects of weather forecasting with not too much difficulty. This remained the unwritten policy of the Meteorological Division for many years.

[1] See the author's *The Beginnings of Canadian Meteorology* (Toronto: ECW Press, 1991).

[2] Bernhard Haurwitz, "Meteorology in the 20th Century," Bulletin of the American Meteorological Society, vol. 66, no. 5, (May 1985), p. 501.

[3] Ralph Jewell, "The Bergen School of Meteorology," *Research in Norway*, 1979, pp. 1–8.

[4] For an extensive account of the 1928–1932 airmail weather service and of activities of the Meteorological Service in the years leading up to 1936 see the author's *Forecasts for Flying: Meteorology in Canada 1918–1939* (Toronto: ECW Press, 1996).

[5] Ibid., pp. 155–157.

[6] Dr. Cody, President of the University of Toronto, in an address of welcome to the joint meteorological meeting in Toronto, August 1939, *Quarterly Journal of the Royal Meteorological Society*, vol. 66 supplement, (1940), p. 9.

[7] Ralph E. Huschke, *Glossary of Meteorology* (Boston: American Meteorological Society, 1959) pp. 237–238.

[8] Forecasting for transcontinental and transatlantic aviation is covered in chapters 13 and 14 of *Forecasts for Flying*.

[9] Carl A. Christie, *Ocean Bridge: The History of* RAF *Ferry Command* (Toronto: University of Toronto Press, 1995), p. 15.

[10] AES, File 5930-40-0, Croil to director, 22 January 1937.

[11] Ibid., Commanding Officer, No. 5 (CR) Squadron to Patterson, September 1937.

[12] Royal Meteorological Society, Canadian Branch, *Quarterly Journal of the Royal Meteorological Society*, Supplement, vol. 66, (1940).

CHAPTER 3

THE EARLY WAR WEEKS

John Patterson, Controller of the Meteorological Division at Head Office in Toronto, was not totally unprepared when war broke out on Labour Day weekend, 1939. For years he and his deputy, Andrew Thomson, had known that, should another war occur, meteorology, and especially aviation meteorology, would be considerably more important to Canada's war effort than had been the case in the 1914–1918 Great War. But Patterson, or anyone else, had no idea of the task ahead of them to recruit and train metmen and meteorologists to meet the needs of the Royal Canadian Air Force.

The Meteorological Division had been building a very competent cadre of trained, young meteorologists and assistants to serve civil aviation. It was thought that the requirements from military authorities for aviation meteorology could be met by moving a few meteorological assistants from the civilian offices to military air fields and filling the positions these men vacated with temporary help. As it turned out, each forecasted need for metmen in the early war months was exceeded as soon as it was made. Over the next few years, nearly a tenfold expansion in the Meteorological Division's professional and technical staff became necessary.

Two noteworthy changes had been made in operational meteorology just before the war began. On July 1, 1939, by international agreement, the millibar was introduced as the unit of pressure to replace the height of a column of mercury in inches. On the same day, the international numeric code for the transmission of synoptic weather observations became mandatory in Canada and the United States. Unchanged, however, was the unique sequence code for hourly observations required by the North American commercial aviation community.

Immediately the war began a new weather forecast office was requested to serve the air force and navy on the East Coast. An office was established at Halifax and within days began to issue forecasts. Planning was commenced on a system to provide meteorological services to a Royal Canadian Air Force flying training program. A complete ban on broadcasting and publishing weather information in eastern Canada came into effect with the outbreak of war, although within weeks some dissemination of weather information was again allowed.

New and restrictive data exchange procedures with the United States Weather Bureau had to be negotiated and put into effect. Within days the coding of all weather information transmitted by wireless stations began and lasted until near the end of the war. However, most ordinary routine meteorological activities were maintained in the fall of 1939, especially in central and western Canada, those areas of the country furthest removed from the European hostilities.

A NEW FORECAST OFFICE

On September 3, 1939, Britain declared war on Germany. Within a few days, even before Canada officially entered the war a week later, the Department of National Defence asked the Department of Transport to establish a weather forecast office in Halifax to provide service for the Royal Canadian Air Force and the Royal Canadian Navy. The RCAF's Eastern Air Command was responsible for the protection of eastern Quebec, the Maritime provinces, Newfoundland, and coastal areas as far north as Labrador while the navy was concerned with the harbours and coastal waters as well as anti-submarine protection for convoys. At about the same time, the High Commissioner for the United Kingdom requested Canada provide a meteorological service for the Royal Navy operating in the Atlantic west of 30 degrees west longitude.

John Patterson had anticipated such a request from the Armed Services and he immediately announced that the new office would be located at the headquarters of Eastern Air Command in downtown Halifax rather than at the Dartmouth airport as some had expected. On September 6, J.R.H. Noble, then officer-in-charge of the Montreal St. Hubert District Aviation Forecast Office, was dispatched to Halifax to establish the required office and to organize the work.[13]

Meteorological activities in Halifax before the war were limited to a traditional telegraphic weather reporting station at the Halifax Citadel and part-time hourly observations by the RCAF at Dartmouth airport. Noble negotiated with the Department of Public Works for office space, for the necessary renovations, for furniture, and for the other physical arrangements necessary in establishing an office. This was largely accomplished within two weeks. An expanded meteorological observing program was planned with additional stations in Nova Scotia, and secret codes had to be introduced for the wireless transmissions of weather information in the coastal areas. A dedicated teletype circuit from Montreal to Moncton, which was being installed along the transcontinental airway to allow the provision of services to TCA, was extended to Halifax within a week.

In the second week of September, experienced staff began to arrive in the new office. Five meteorological assistants to act as observers and map plotters, four meteorologists to do the weather forecasting, and four teletype operators were dispatched to Halifax from the Montreal St. Hubert and Toronto Malton District Aviation Forecast Offices. But getting the meteorological communications system to work turned out to be rather more difficult and time consuming as the largely inexperienced operators left at St. Hubert were unable to relay a satisfactory volume of meteorological data to the new forecast office for some time. Further, the absence of suitable landlines in Newfoundland meant that contact with Gander had to be made by radio with all messages transmitted in secret code. Despite the data limitations, the Halifax forecast office was able to begin operations on September 11, but it would be the end of the month before a minimum acceptable level of communications was reached to allow the office to function properly.

Heading a civilian-staffed meteorological office operating within a military environment, Noble faced possible serious management problems at Halifax. This was averted when the Departments of National Defence and Transport accepted a simple and practical working agreement. The staff at the forecast centre would come under the immediate control of a senior RCAF officer but the actual meteorological operations were to be controlled by Noble's superior, John Patterson, the civilian head of the Meteorological Division in Toronto.

Within weeks, the RCAF, realizing their need for help in interpreting the forecasts and briefing the pilots, asked for additional civilian meteorological personnel for posting to the Home War Establishment bases in Sydney and Yarmouth, Nova Scotia. However, the RCAF had to be told that it would be some time before meteorologists would become available and so the air force posted airmen with some meteorological training and briefing experience to those airports. There was a civilian observer at Shediac, New Brunswick, to provide observations for the Newfoundland to New York leg of the transatlantic flights and, in October 1939, he was moved to Saint John airport where he commenced hourly observations. The radio range operators at Moncton and Blissville airports on the transcontinental airway also began weather reporting that month.

But the lack of nearby weather reporting stations was a very serious handicap for local forecasting at the new Halifax office. To overcome this, plans were made to establish new stations on either side of Halifax at Ecum Secum and Liverpool. Unfortunately, there were major delays in setting up these stations and the observers were not in position to report until July 1940. With no telegraph or teletype circuit available, both stations used the telephone to report to Halifax and this method of communication continued throughout the war.

As a civilian organization, the Meteorological Division encountered serious obstacles in getting the necessary meteorological system operating in Atlantic Canada during the first few months of the war, obstacles that were at first not appreciated by the military who demanded more people and more services. As already noted, meteorologists, meteorological assistants, and teletype operators were transferred to Halifax from central Canada but more were urgently needed. While meteorologists were being trained in Toronto for the entire system as rapidly as possible, weather observers and teletype operators had to be recruited and trained in Atlantic Canada according to peacetime Civil Service Commission staffing rules despite the urgent need for the people. It was thus necessary to hold competitions and establish eligible lists involving prescribed waiting periods before any positions could be filled. As admirable and essential as this system was in peacetime, it was exceedingly slow and cumbersome during the early weeks of the war and it retarded by many months the establishment of the full meteorological system required by the military authorities on the East Coast.

SERVICE COMMENCED

The Royal Canadian Navy (RCN) required forecasts for the coastal areas around Nova Scotia and in the Gulf of St. Lawrence as well as special weather information for the dockyard area whenever convoys were arriving or departing. These were provided and, in time, the chart room at the Halifax Dockyards was connected to the Moncton to Halifax meteorological teletype circuit so that all weather information and forecasts would be available to the naval officers. Later, direct telephone connections were maintained for consultation on any meteorological problems experienced by the RCN and visits to the dockyards by meteorologists were arranged, especially on days before convoys were about to leave for overseas.

The Dartmouth-based RCAF squadrons required flying-weather forecasts for their operations along the East Coast from Belle Isle south to Yarmouth, including the Bay of Fundy, and eastward over the Atlantic to Sable Island. Four forecasts a day for all areas were requested and this forecast service began on September 30. Pre-dawn forecasts for operational patrols were valid from dawn to dusk while a general forecast for the next day assisted the senior officers in planning operations. Weather forecasts and other meteorological information were sent by courier to the Dartmouth base and from Halifax to the bases at Sydney, Yarmouth, and Saint John using the

RCAF teletype system. The RCAF was anxious to have forecasting staff at these four bases but this was not possible until the Meteorological Division could train more meteorological officers.

The Halifax forecast office encountered unexpected difficulties when it was soon realized that the senior RCAF officers were almost completely unfamiliar with the terminology used in weather observations and forecasts. Consequently, at first, all weather forecasts and reports of current weather had to be expressed in plain language without using symbols, abbreviations, and codes. Because the officers maintained they had no time for classroom instruction or visits to the forecast operations room, an educational program was developed by Noble in which circulars were written and issued on clouds, frontal systems, and other topics which might be studied by the officers when off duty. At the Dartmouth air base the Halifax meteorologists organized a refresher course for the active pilots who, as a rule, had some earlier meteorological training.

NEWFOUNDLAND

Newfoundland was not yet part of Canada in 1939 and so the organization of meteorological activities there was somewhat different from that in the Canadian provinces. For decades, weather observations had been sent to Toronto from a few Newfoundland stations and public weather forecasts were sent back to the island. Then, in 1937, for the experimental transatlantic flying boat program, the Canadian Meteorological Division became responsible for an operational meteorological system with a forecast centre at Botwood, which then moved to Gander in December 1938. The office consisted of meteorologist-in-charge Pat McTaggart-Cowan, three additional meteorologists, Hugh Bindon, Gib Henry, and Keith McLeod and eleven assistants. There was also an office in St. John's where a technician (James Kirk) supervised the Newfoundland observing stations — nine regular synoptic observing stations and more than two dozen supplementary observing stations where partial observations were taken daily and telephoned to Gander.

These meteorological facilities were used to provide services for the transatlantic British and American flying boat flights, which had begun in the summer of 1937. From the end of June 1939 until mid-September, Pan American Airways had made twenty-six flights carrying mail and passengers, and Imperial Airways sixteen mail and express flights while a few trips were made by American Express Airlines and Air France TransAtlantique.

When war came, the British flights were discontinued but, with the United States a neutral power, the American flights continued for some weeks although all broadcast weather information had to be in secret code and the ships at sea stopped transmitting weather observations. At the end of the season, in October 1939, the British and Canadian governments issued instructions to shut down operations at Gander and mine the runways in case Newfoundland and the airport should fall into enemy hands.

However, the senior meteorological, communications, and operational officers at the airport were certain that flying the Atlantic with land-based aircraft would not only become possible but might become critical to the war effort, so they delayed action to close the airport. Then, by the spring of 1940, "there was a sudden awakening."[14] The RCAF now realized that their aircraft could not cover the entire Newfoundland coast from the Dartmouth base and they wished to use Gander. Also, governments soon realized that the loss through enemy action of ocean freighters carrying new American-made aircraft to Britain was so great that an attempt had to be made to fly the aircraft over the Atlantic. Gander airport and the meteorological facilities on the island were now becoming of great value to the war effort.

BLACKOUTS AND CODES

Early in 1937, the British Meteorological Office expressed concern to the Meteorological Division that, should war break out, it was almost certain that the enemy would intercept wireless messages containing weather information. Accordingly, secret codes were prepared and issued in sealed envelopes to all northern isolated stations where radio was used to transmit weather observations. These envelopes were to be opened only if war occurred. Also, discussions were held and decisions made regarding the exchange of data with the Americans should Canada and Britain be at war while the United States remained neutral.

It was fortunate that Canada and the United States had adopted the International Meteorological Organization's numerical code for the transmission of synoptic observations on July 1, 1939, since it would have been exceedingly difficult to develop secret codes based on the old word code.

Immediately after September 3, when the United Kingdom declared war on Germany, the Canadian Board of Censorship took steps to suppress meteorological information to prevent it from falling into the hands of the enemy. The Meteorological Division was requested to —

a) introduce secret codes for the radio transmission of all Canadian weather information except on the Pacific coast,
b) stop the broadcast of weather information over the radio in plain language,
c) stop broadcasting storm warnings,
d) radically curtail the distribution of weather information by telegraph,
e) stop the publication of weather forecasts in the newspapers of central and eastern Canada and
f) cease the transmission by radio of weather observations from ships at sea.

Meteorologists could accept these restrictions since they knew that even a little knowledge of the weather over Canada would allow enemy meteorologists to visualize the existing synoptic weather situation and allow them to make forecasts which would aid their Armed Forces in the Atlantic and even in Europe.

The Canadian radio-reporting observing stations were instructed to begin using the secret code on September 3, 1939. This necessitated decoding the messages in the South, and this was done at Toronto for the messages from the northern stations and at Halifax for those from Newfoundland. The work at Halifax was soon transferred to St. Hubert airport where more appropriate facilities were available and because it became the base for transatlantic meteorological operations. At the Toronto headquarters it was soon found necessary to set up a rather large code and cipher office so that the

District Aviation Forecast Offices and other inland meteorological offices would not be required to undertake any coding and decoding work. Decoded, the information could be sent over the teletype "in the clear," that is, using regular meteorological codes.

Also, it was found necessary to maintain a coding staff in Newfoundland to handle the coding of observations and forecasts for transmission to Britain and to decode any radioed observations not received at Toronto. The Newfoundland staff also handled the coding of weather information sent to the United States for use with the American transatlantic flights. Some of the coding work was later discontinued when landlines were constructed in Newfoundland allowing the establishment of a teletype system which carried observations and forecasts received by cable from Nova Scotia. The coding and decoding of all weather data and information broadcast by wireless in Canada proved to be a burdensome task which had to be continued until nearly the end of the war.

PUBLIC WEATHER

Early in September 1939, Canadian newspapers and radio stations were instructed by the federal government to discontinue the publication and broadcasting of weather reports and forecasts. This deprived the public of a popular, useful, and customary service and was especially inconvenient to coastal fishermen, ship owners, and mariners at Halifax and at other Atlantic ports who depended on daily weather information in their work. The ban remained although, within a few weeks, the publication of local forecasts in the newspapers was again allowed but no mention was to be made of barometric pressure or wind direction.

Although the practice of hoisting storm warnings was to be continued, the lack of radio broadcasts was particularly hard on the mariners. Naval authorities were allowed to broadcast the storm warning signals on occasions when enemy activity in the Atlantic was such that no damage would likely come from this break in the blackout of weather information.

At this time all public weather forecasts for the Maritime provinces were issued from Toronto but, in the early summer of 1940, the new

Halifax forecast office began to prepare forecasts for the Annapolis Valley fruit growers. These forecasts were sent to the telegraph companies by teletype and upon identification were available by telephone to the fruit growers. In Newfoundland, the Gander forecast office had already taken over the provision of storm warnings, public weather forecasts, and forest fire hazard forecasting for the island.

The publication of weather summaries and stories from other parts of the continent was not allowed. However, apart from these special circumstances in Atlantic Canada, the Meteorological Division maintained a normal provision of public weather services in the rest of Canada during the early weeks of the war although a ban on radio plain language weather reports and forecasts did exist for some weeks. General public weather forecasts were issued twice daily for twenty-eight regions throughout the country from Toronto and Victoria in which the probable weather for the next twenty-four to forty-eight hours was given. There were special forecasts for forest fire hazards, and in spring and summer, forecasts for spraying operations and to warn of frost in the fruit belts. The hoisting of storm warnings continued when necessary for fishermen and mariners on the coastal areas and the Great Lakes.

The preparation and publication of the daily *Weather Map* at the Toronto headquarters became a wartime casualty. At first, the *Weather Map* with its regional forecasts was prepared and printed daily and then held for a week before distribution. However, this considerably diminished its usefulness and a decision was made to discontinue the *Weather Map* for the duration of the war. In the immediate post-war years, the publication of a map at Toronto was judged to be unnecessary when many newspapers began to publish a weather map each day.

OBSERVATIONS

The maintenance of the national weather observing system in wartime was absolutely necessary for the preparation of forecasts for both the general public and aviation interests. Consequently, this program was continued without interruption after the war began. In September 1939, there were about thirty new airways stations reporting both hourly and synoptic observations as well as another 110 or so part-time observers who reported at least twice daily by telegraph. In addition, there were another 850 or so climate observing stations that reported monthly by mail. Most of these stations were administered and monitored from the Toronto headquarters.

The observing station inspection program, where in pre-Depression years an attempt had been made to have a weather forecaster visit each observing station once every two years or so, had faltered in the early 1930s when budgets had been slashed. In fact, during the late-1930s, there had been little or no time for meteorologists to resume a normal inspection program. The work of instructing new observers at the airways stations had fallen on existing staff at each station and on some new meteorologists as they travelled between postings.

The requirement for aviation meteorology services for Trans-Canada Airlines increased as passenger and mail services were expanded. The main TCA route was extended from Montreal to Moncton early in 1940 and to the Dartmouth airport at Halifax in April 1941. The Halifax office was not in direct communication with TCA and besides it was exceedingly busy serving the air force and the navy, so the Montreal DAFO provided landing forecasts at Moncton and subsequently Dartmouth airports until a District Aviation Forecast Office opened at Moncton in 1942.

There were difficulties in providing weather information to the TCA aircraft in flight by radio in eastern Canada early in the war. The information had to be in secret code, which was changed every day, and this did not allow the captains time to decode the messages which could be of significant value to them. Consequently, changes were made to simplify the coding of vital cloud and visibility information.

EXCHANGE WITH THE UNITED STATES

Since 1872 there had been a traditional free and unhindered exchange of meteorological observations between Canada and the United States. By 1939, each national Service had its own dedicated teletype circuits and these were linked in the East between Toronto and Buffalo and in the West between Vancouver and Seattle. Six-hourly synoptic weather observations, hourly observations or "sequences," radiosonde observations, and pilot balloon observations were exchanged according to predetermined schedules arranged to meet the needs for analysis and forecasting and for the provision of relevant information to pilots. Canadian weather data were broadcast by the Americans from Arlington to ships and aircraft over the Atlantic and to Europe. Canadian hourly weather observations were broadcast from radio range stations in the United States just south of the international border and, similarly, American observations were broadcast from Canadian radio range stations.

When the war began, the Department of National Defence prohibited the broadcast of weather information over the radio in plain language and ordered the introduction of secret codes for the transmission of all data except on the West Coast. The Americans were unhappy with this since they wished to continue their daily broadcasts, especially from Arlington to their aircraft over the Atlantic Ocean and their ships at sea. To solve this dilemma, John Patterson went to Washington early in September 1939 and reached an agreement with the Americans. It was acknowledged that the Arlington broadcasts could not be suppressed but it was agreed that all eastern Canadian weather observations would be omitted.

The radio transmission of weather information required by the Americans for flying from Newfoundland would continue using a special secret code. Canadian observations would be provided by teletype to the Weather Bureau and the U.S. military for their own internal use. The Americans then set up a special forecast centre to map and analyze all available data and prepare forecasts, which were then encoded and distributed as encrypted wireless and teletype messages.[15]

A MILITARY AIR TRAINING PROGRAM

During the Great War the Royal Flying Corps had recruited Canadians and trained them at flying training schools in Canada. Over the fifteen or so years following the war, nearly 1,000 Canadians had gone to England to become members of the Royal Air Force and half of these were trained as aircrew. In 1935 the British government proposed a "Trained in Canada Scheme" under which the RCAF would train a certain small number of Canadian pilots each year for the RAF.

The Canadian government accepted the proposal, a course of fifteen was trained in 1938 and a second course was in training when the war began.[16]

As the war threat increased in 1938 and 1939 the British government sought to have more pilots trained in Canada and new negotiations were commenced. The matter became very political; the Canadian government insisted that training in Canada be controlled by Canadians and that the additional trainees be men from Britain and not from Canada. During the negotiations which followed, the RAF and the RCAF agreed on many non-political training issues such as the necessity for a common training syllabus and that pilot training should be accomplished in three stages — elementary, service, and operational.[17]

Discussions intensified after the war began and Britain asked Canada to train 8,000 pilots a year for the Royal Air Force. The Canadian government was dumbfounded by the number suggested but the possibility of staging a major air training program, and having it considered Canada's major war contribution, became intriguing to a government hesitant about attempting to raise a large infantry army for service overseas. In September 1939, Vincent Massey, Canada's High Commissioner in London, and his Australian counterpart, Stanley Bruce, began to advocate a commonwealth air training program under which manpower resources would be pooled, elementary training be done in several commonwealth countries, and the advanced or service training be done in Canada. Graduates would then join their country's squadrons with the RAF in Britain. Accord on general principles was reached amongst representatives of Britain, Australia, New Zealand, and Canada and a meeting to complete a formal agreement was arranged for mid-October in Ottawa.

At these meetings, Canada, as the proposed host for the major portion of the British Commonwealth Air Training Plan (BCATP), initially part of an overall Empire Air Training Plan, demanded and was promised control of the bases in Canada and of the service personnel involved. It proved to be more difficult to reach agreement on the financial and political issues — how the costs would be shared, when funds would be transferred, where the training aircraft would be purchased, and the international balance of payments relationships with the United States. These matters were discussed at length before final agreement was reached.

In the end there were two political points on which Prime Minister Mackenzie King insisted. The first was his stipulation that a statement be issued by the United Kingdom government that "the Plan would provide for more effective assistance . . . than any other form of military co-operation which Canada could give." The second point was King's demand that a statement be published that "the Canadian pupils when passing out from the training scheme will be incorporated in, or organized as, units of the Royal Canadian Air Force in the field." King won both points at the last moment and Britain and Canada signed the BCATP agreement on December 17, 1939. The Australia and New Zealand delegates had gone home by that time but they signed at a later date.[18]

There is no evidence that the national meteorological services of Australia and New Zealand were ever involved in planning the necessary meteorological services for the BCATP. However, while the high-level political negotiations were proceeding, correspondence had begun between the Canadian Meteorological Division and the British Meteorological Office. Subsequently, the London Meteorological Office provided the Toronto headquarters with full information on training syllabi, instruction manuals, observing and forecasting forms, and other technical information used in giving meteorological training to Royal Air Force aircrew. The Meteorological Office suggested sending a representative to assist in organizing the BCATP meteorological training work in Canada but the shortage of meteorologists in Britain eventually prevented this from being done. Similarly, Patterson claimed he was too short of personnel to be able to send a representative to the United Kingdom to work out an organization plan for meteorological aspects of the BCATP. (Later, in 1940, the United Kingdom government did send C.W.G. Daking, a Civil Officer (Meteorology) to Ottawa to become a member of the U.K. Air Liaison Mission.)

The story of the Meteorological Division's efforts in organizing, staffing, and operating a meteorological training program for the British Commonwealth Air Training Plan will be covered in later chapters.

[13] AES, Archives, J.R.H. Noble, oral interview with David Phillips, January 12, 1984.
[14] Ibid., P.D. McTaggart-Cowan, oral interview with David Phillips, October 5, 1983.
[15] In *American Weather Warriors* (Texas A&M University Press), Bates and Fuller describe how the Americans solved this security problem.
[16] F.J. Hatch, *Aerodrome of Democracy: Canada and the British Commonwealth Air Training Plan* (Ottawa: Department of National Defence, 1983), p. 6.
[17] Ibid., p. 14.
[18] Ibid., pp. 13–26.

CHAPTER 4

MAKING METMEN

During World War II the Department of Transport's Meteorological Division hired hundreds of university graduates and trained them to serve as Meteorological Officers with the Royal Canadian Air Force. On the RCAF stations, the flying instructors and the student pilots and navigators referred to these men as the "metmen." The Civil Service Commission officially classified the metmen as Meteorological Assistants Grade 3 since they lacked the graduate degree in physics (meteorology) which was necessary for classification as a "meteorologist."

It was almost completely a male profession as apparently no attempt was made to recruit women until 1943. In two courses in 1943–44 nine women were trained as "metmen" and three of those later took an advanced course and became meteorologists. The meteorologists were posted to forecast offices but there is no record of "women metmen" at RCAF stations. However, there were other women engaged in wartime meteorology with the air force, members of the RCAF's Women's Division, who assisted the metmen in the station meteorological sections. These airwomen began to replace airmen as observers and map plotters in 1942 and were referred to as the "WDs."

METMEN IN WARTIME

THE CONCEPT

During the summer of 1940, as plans for the training of aircrew under the British Commonwealth Air Training Plan kept expanding, it became evident that the RCAF would require many more meteorological officers than the twenty-seven men originally requested in March. This was because the air force soon found that more than one meteorological officer would be required at each school, and besides, the airfields were being constructed and schools opened at a much faster rate than planned. Further, the Commonwealth governments now realized that as the European war situation deteriorated, aircrews for the Allied Forces were needed in abundance and this need would continue for some time.

That summer, Patterson and his assistants in the Meteorological Division found that the normal University of Toronto graduate course would not produce the number of meteorologists required quickly enough nor did metmen need that level of training for the work at the RCAF stations. So, as well as continuing the MA course in academic year 1940–41 to produce more meteorologists, it was decided to hold two successive short intensive courses to produce metmen for service with the RCAF. Each course would be of fifteen weeks duration and would train about fifteen "dependent forecasters" in contrast to the MA graduates who were considered to be "independent forecasters." The first course would begin in November 1940.

While the Civil Service Commission classified the MA graduates as Meteorologists Grade 1, it refused to accept the term "assistant meteorologist" for those who completed the new meteorological officers' course and so they were classified as "Meteorological Assistants Grade 3." In wartime temporary positions, most would be seconded to the RCAF to provide meteorological services in the British Commonwealth Air Training Plan. After these men successfully completed the course and were posted to RCAF stations, the air force personnel soon labelled them "metmen."

DUTIES AND QUALIFICATIONS

For recruiting purposes, the Meteorological Division provided to the Civil Service Commission a simple set of duties and qualifications for the positions:

> *Duties*: After a period of instruction to be attached to an Air Training School to give instruction in Elementary Meteorology or to analyze weather maps, interpret forecasts and allied work. *Qualifications*: Graduation with specialization in Mathematics and Physics from a university of recognized standing or at least an honour mathematics and physics course equivalent to that at the University of Toronto, accuracy, speed, neatness, good eyesight, reliability, good address and teaching ability.

In early 1941 a "duties" circular was prepared and made available to prospective recruits and in late 1942 another circular was issued for those considering whether or not to apply to become a metman.[19]

As with many positions, the advertised duties did not always include all the actual duties handled by the metmen during the war. Most metmen did spend the duration of the war at RCAF stations where British Commonwealth Air Training Plan aircrew students received a significant portion of their training. Some were posted to RCAF stations on the East and West Coasts to brief aircrews engaged in operations and/or assist the meteorologists at those bases. In "ground school" at the training stations the metmen gave courses of meteorological instruction and for flying training they gave the instructors and students forecasts and weather briefings. Although classed as dependent forecasters, the metmen often found it necessary to prepare their own local weather forecasts as guidance from the regional DAFO was received too late to be of much assistance in allowing them to meet the demands of the air force.

RECRUITING

The need for metmen increased so rapidly in 1940 that by the time the Civil Service Commission advertisement for them appeared in the newspapers that fall, the number required was increased to ninety men who "will shortly be required in connection with the BCATP." A second advertisement in March 1941 indicated that not only would applications from graduates be accepted but applications from students who had successfully completed three years of an honour mathematics and physics course would also be accepted. For some time contacts with prospective applicants were by letter since the first information circular on the subject was not issued until 1942.[20] A copy of part of a personal letter from John Patterson to a potential applicant appears as Appendix 1.

At first there was, however, considerable bureaucratic delay in filling the positions since the normal Civil Service Commission (CSC) rules had to be followed. There was a pause of a matter of weeks for possible applicants to respond after the competition was announced and further delays were necessary to allow time for holding a rating board and creating an eligible list. And there was more waiting for possible appeals, and only then were job offers mailed out.

Another factor that caused delay was the CSC position that, because honour university graduates were involved, the salary for these positions should be $1,800 rather than the $1,620 then paid to those already in the service at the Meteorological Assistant Grade 3 level. It was pointed out, however, that if this was done many of the current Grade 3s, who were also honour university graduates, would apply and be accepted which would necessitate more competitions to fill the jobs they would leave. Accordingly, it was agreed that all grade 3s should receive a salary of $1,620 and, in compensation, salaries for the new entrants should begin once training began rather than after completion of the course.

The first intensive course of nineteen candidates, which began in mid-November 1940, attracted a number of excellent would-be metmen. Most were to be subsequently accepted for more training, later became meteorologists and make meteorology their post-war careers. They were mostly recent university graduates who had begun careers in business and industry or who were in graduate school. Consideration had been given to seeking bilingual candidates but since the RCAF had not made a decision on this matter in connection with the navigation course, the other largely academic course of ground school instruction, no action was taken.

By 1941, after the first two courses, the supply of graduates with work experience was largely diminished but the need for metmen became so urgent that the Meteorological Division began offering positions to qualified candidates without waiting for the bureaucratic steps. (The official documents came through weeks after a successful applicant was on course.) Recruits were solicited directly from the universities where mathematics and physics professors were advised periodically of the need for graduates to train as metmen, and such information was spread by word of mouth though student circles or by notes posted on bulletin boards. This brought applications from fresh graduates without other work experience although quite a few high school teachers were recruited and joined after the 1941–42 school year. Males of draftable age realized that they had to contribute to the war effort and at that time could choose between volunteering for one of the Armed Services and applying for the course in meteorology.

The recruiting situation in the spring of 1941 at the University of Western Ontario might be typical of what happened at other Canadian universities at that time. At that institution there were about a dozen graduating male students in fourth year honour mathematics and physics. Enrolled in the Canadian Officers Training Corps, many were taking artillery training with the prospect of becoming officers in the Royal Canadian Artillery on graduation. But, by the summer of 1941, the graduates in the physics option chose to go into radar work with the Royal Canadian Navy and graduates in the mathematics option chose to become metmen. In addition, four men who had successfully completed three years of honour mathematics and physics at Western were also accepted for training as metmen.

As the war progressed and the manpower shortage became acute, the Meteorological Division was forced to begin accepting university graduates with only a few, if any, mathematics and physics credits. Part of this was due to a Wartime Bureau of Technical Personnel ruling in 1942 which required all 1943 graduates to register and file a declaration indicating which of the Armed Services they wished to join as technical officers. Names of those not selected were given to the Meteorological Division and more than sixty desperately needed men and several women were hired for training as metmen that year. Accordingly, the course that began in May 1943 had to be divided in two as those with little mathematics background were given an extra month's training.

More than fifty years later it appears unusual that the Meteorological Division hired so few women to train as "metmen." While female secretaries and typists were employed before the war there had never been any women professionals in meteorology. It was perhaps the success of the airwomen (the WDs) in the meteorological sections at the RCAF stations and the difficulty in recruiting men that impelled John Patterson and his Head Office staff to break with tradition and recruit qualified women for the Intensive Course in Meteorology. Note has been made previously that nine women were recruited for courses in 1943–44 and that three did exceptionally well and were invited to take the advanced course and became forecasters. Unfortunately, only two of the nine remained in meteorology after the war and neither as a metman or meteorologist.

In 1943 the government recognized the essentialness of the meteorological work in that all metmen and meteorologists were "frozen" in their positions for the duration of the war. There was an attempt to take the metmen into the RCAF, and when this did not happen many metmen became dissatisfied and attempted to enlist in one of the Armed Services but were prevented from doing so. There were some exceptions: an agreement was reached between the Department of Transport and the Royal Canadian Navy under which a few metmen were released to take commissions for meteorological work in the navy.

The last short course was held in 1944. In all, there were a total of twelve such courses and 373 new metmen were graduated. However, not all remained in a "dependent forecaster" metman position with the RCAF. Soon after the MA meteorology course was completed early in 1941 it was realized that many more meteorologists would be needed. So another wartime course, the Advanced Course in Meteorology, was developed for the best graduates of the short course to upgrade their training and skills to the level of those who had taken previous MA courses. The first advanced course of fifteen men began in October 1941 and ran for fifteen weeks. The need continued and a total of seven advanced courses were staged with a total of 120 graduates by the summer of 1946.

It must be noted that not all those who were trained to become metmen were recruited during the war years. There was another source, a source within the Meteorological Division itself. When the DAFOs and observing stations on the transcontinental airway were staffed during the late stages of the economic Depression, from 1937 to 1940, several of those hired were university graduates. These men had been unable to find other jobs that would fully use their education and training. Within the Meteorological Division their capabilities were soon recognized. During the first year or so of the BCATP nearly thirty of these men without formal meteorological training were posted to RCAF schools to serve as metmen until trained metmen became available. When replaced, over the next year, they were brought back to Toronto and given the training they had missed. Later, many of them were invited to take the advanced course and become meteorologists; in fact, a few of the self-taught metmen were invited to take the advanced course directly without first taking the short course. About twenty of these men who began their careers as observers made meteorology their peacetime career and were to lead the postwar expansion of meteorology in Canada.

TRAINING

Each Intensive Course in Meteorology ran for fifteen weeks. No more than six or so instructors, usually meteorologists with forecasting experience, were responsible for the training program. It was a demanding job especially as one course usually followed directly after another and on occasion an advanced course was held simultaneously with a short course. Classroom space for instruction was always a problem. There were no lecture rooms and little or no space for map laboratories at the Head Office building, 315 Bloor Street West, although laboratory sessions in some courses were held in a large first floor office and lectures were given in a confined tower room on occasion. But, for most courses, space had to be found in other buildings near Head Office.

For example, with the fourth short course which ran from mid-October 1941 to the end of January 1942, the trainees moved at least once every day from building to building carrying their books as well as their practice weather maps. For that particular course most lectures were held and map work done in a room at the Royal Ontario Museum while at other times the basement of the University of Toronto's Economics Building (the old McMaster University building and now the Faculty of Music building beside Varsity Stadium) was used. On occasions the course attended lectures in the 315 Bloor Street building and in other parts of the University of Toronto's McMaster building.

Even for most honours mathematics and physics graduates the short course proved to be rigorous and called for much evening study. The course week was five and a half days long and each day was scheduled from 8:30 a.m. to 5 p.m. and until noon on Saturdays. The course was, in a sense, the first half of the Toronto Master of Arts course in physics (meteorology), with practical map work every afternoon. The map laboratories took more hours than one might think necessary since in those days, without copying machines or computers, the trainees had to plot their own maps from signals (observations) called aloud and that took more than half an hour of each daily laboratory session. There were three lecture periods most mornings and sometimes a lecture was inserted into the afternoon program. The syllabuses of the wartime intensive and advanced training courses are given in Appendix 2.

The instructors were mostly young meteorologists who had graduated from the MA course and then acquired a few years experience at District Aviation Forecast Offices. For example, the meteorologist primarily responsible for the fourth metman course in 1941–42 was Jim Leaver who had had operational experience at the Montreal St. Hubert DAFO and at the Trenton RCAF station. He lectured in synoptic meteorology, supervised the map laboratory sessions each afternoon, and, towards the end of the course, presided over the practice lecture sessions. Des Kennedy, who had had practical experience at the Winnipeg DAFO, was primarily responsible for the first advanced course which was being conducted at the same time, and he assisted with the short course laboratory sessions. (The map laboratories of the advanced course were held in the mornings and those of the short course in the afternoons.)

Leaver gave an elementary course in meteorology early in the training period that was designed to tie in with beginning the practical map work. The theoretical course lectures in dynamic and synoptic meteorology were given by Dick Longley, who had forecasted at Montreal St. Hubert, and by Wendell Hewson, an early MA graduate who had subsequently obtained a Ph.D. in meteorology at the University of

London. Later, two other meteorologists who were forecasting were brought back to Toronto to assist in the training program, Al Crocker in 1942 and Clarence Penner at the end of the war.

W.E.K. (Bill) Middleton of the Instrument Section usually gave a course in meteorological instruments which was of practical value and would prove useful later when the metmen were supervising airmen observers on RCAF stations. A weekly lecture in climatology from Griffith Taylor, an eminent geography professor at the University of Toronto, was somewhat less than satisfactory from the trainees' viewpoint as it failed to relate to other aspects of the course or to the work for which they were preparing. The dynamic meteorology course, which included thermodynamics, was particularly rigorous. Practice lecture sessions were scheduled since the students would be expected to give courses in meteorology to student aircrew when posted to BCATP stations.

For many students the map laboratories proved to be the most interesting and absorbing. By being forced to plot their own weather maps, the students quickly learned the location of the weather observing stations and how to plot the information around each station circle in the standard prescribed manner. As the students undertook their map analysis, an instructor or two and perhaps a trained metman brought in for this task, circulated behind the desks or tables and observed each trainee's individual analysis of the meteorological situation and offered criticism and advice. Historical map series employed were selected to illustrate typical or classical meteorological situations and towards the end of the course current maps were used.

Most graduates of the first three short courses spent two or three weeks at a specific regional DAFO before proceeding to their RCAF postings. This temporary posting served to introduce each new metman to practical weather forecast operations and to the forecasters whose area forecasts he would be using as guidance after he arrived at his RCAF station. The posting was also to alert him to any peculiarities of the weather in their region. However, by the time the fourth course graduated, in early 1942, the demand for metmen by the RCAF was such that the graduates were posted directly to the RCAF training stations and took forecasting and lecturing shifts from the day of their arrival. Tables showing the graduates of the wartime intensive courses and other metmen are shown in Appendix 3.

POSTING

The metmen were usually given their postings after the final examinations, a day or two from the end of the course. The individual postings were made by the deputy head of the service, Andrew Thomson, and were based on consultations with the instruction staff who had got to know the student-metmen over several weeks. Preferences were requested but the immediate needs of the Division were of prime importance in determining who went where. The system was flexible enough however to usually accommodate a married metman by granting him a posting where it would be possible for him to take his wife and family. This was not possible at RCAF stations in isolated Pacific coast stations and, for a few years, in Newfoundland.

To establish his identity on arrival at a RCAF station guardhouse,

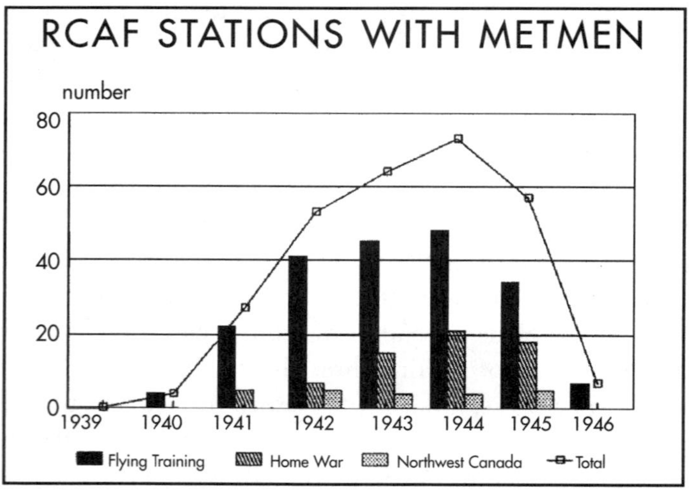

FIGURE 1

The number of RCAF stations at which metmen were stationed each year over the 1939 to 1946 period. Most flying training took place from bases in central Canada and operational flying from bases on the Atlantic and Pacific coasts. Also, some metmen were posted to stations along the Northwest Staging Route which was used primarily by the Armed Forces of the United States.

the metman presented a letter from John Patterson to the Commanding Officer. He was then taken to the station adjutant's office in the control tower where a station pass was issued to him, arrangements made for accommodation, if required, and he was then shown to the meteorological section office.

The metman officer-in-charge, or if he was not present, a fellow metman, then took over and introduced the newcomer to the meteorological section staff and the control tower officers with whom he would closely work. The new metman was then escorted around the station to show him the location of the Officers' Mess and Quarters where he would eat when on station and most likely live, and introduce him to the officers responsible for the ground school where he would lecture. The new metman was usually working within hours and certainly by the day after arrival on a station. Metmen arriving at RCAF operational stations had much the same experience although without student aircrew the stations were usually much smaller than the training stations. A list of the RCAF wartime flying training schools with metmen is given in Appendix 4.

For the inexperienced metman, arrival at his first station was often somewhat intimidating. During the early days of the war, the metman, in civilian clothes, on a station amongst a sea of uniforms, found some RCAF people perplexed in encountering a civilian and perhaps were reluctant to accept him. However, after a year or so, the civilian metmen were completely accepted on air force stations. To the officers, his assignment to the Officers' Mess became proof of status, even if the only other civilians on station were in janitorial or "batman" (valet) work.

AIR FORCE LIVING

To better provide meteorological advice to the RCAF, the Meteorological Division recommended that the metmen live on the station whenever accommodation was available. In the planning stage, the RCAF, although willing to grant officer status to meteorologists, felt that metmen were more the equivalent of non-commissioned officers and thus should live in the sergeant's mess. But, John Patterson argued that as the navigation and some other ground school instructors were university graduates and officers, the metmen must also be given that status. The RCAF reconsidered and agreed that as long as the meteorological personnel were graduates of recognized universities "they will be regarded, for the purposes of status, as officers and as such will use the Officers' Mess and amenities of the stations to which they are attached."[21] At the same time, the few married metmen were permitted to live off the station if they so desired. At some stations, late in the war when overseas veterans were returning to Canada, there was no space available in the Officers' Quarters on some training stations and metmen were forced to live elsewhere.

There were double rooms in the Officers' Quarters and the metman shared a room with a flying instructor or sometimes with the civilian recreation and support officer, either a YMCA or Royal Canadian Legion man. The metman enjoyed the usual "perks" of an officer with the resident batman making his bed, cleaning the room, shining his shoes, and pressing his suit at night. The metmen were charged room and board when living on RCAF stations and those stations operated by civilian companies for the air force. The usual charge was $2 a day, then considerably more than comparable civilian accommodation would cost. Metmen living off the station paid for meals eaten in the mess.

After many complains during the first year, the metmen at RCAF training stations were given a supplement of $30 a month to help cover the cost of accommodation on station. The supplement helped but, from the viewpoint of the metman, the matter was still aggravated by the fact that uniformed personnel did not pay for accommodation on station and that the pay of a metman, $1,620 a year, was substantially less that of the non-flying officers with whom they associated.

In general, the civilian metmen enjoyed a good relationship with the RCAF flying instructors and senior officers. Early in the war, some commanding officers resented having a civilian officer on their station. Often these COs had been pilots in the Great War when meteorological advice was not available or perhaps even necessary. Others, former bush pilots, had flown for years in the North without weather forecasts and neither group believed that metmen were necessary. And, if they were, why were they not in uniform as were the metmen at the

RAF training stations that were opening in Canada? What the senior officers did not understand was that at the RAF training stations in Britain the metmen were civilians and it was only when posted outside the country that they were given commissions and put into uniform.

LINES OF AUTHORITY

The reporting status of the metman at RCAF stations was somewhat confusing at first. Since each had presented a letter of introduction to the Commanding Officer (CO) when arriving, many of the COs believed the metman was directly under his control and that all correspondence dealing with him and the station meteorological section must pass through the CO's office. It was not until March 1941 that an Air Force administrative order was issued clarifying that the Controller of the Meteorological Division had authority over the civilian meteorological staff on all technical matters. Further, the order stated that the civilian meteorologist-in-charge was to be directly attached to the Commanding Officer as his advisor on meteorological matters and administration of the meteorological section should be in accordance with his wishes. RCAF airmen assistants were to be under the metman/meteorological officer for all work connected with the meteorological section. The civilian metman was charged to comply with regulations relating to the maintenance of station discipline.[22]

Rank and authority within a station meteorological section were handled very simply. When a second metman arrived at a station he usually became subordinate to the man who was already there regardless of seniority as a metman. There was no additional compensation for the metman in charge. By the last few months of the war, when there were three to five experienced metmen at some stations, a decision as to who would be in charge was made at Meteorological Division headquarters. The OIC had responsibility for scheduling shifts of his metmen and airmen assistants, for corresponding with Meteorological Division headquarters on all administrative matters, for dealing with the station Commanding Officer, and for handling any difficulties with the airmen assistants.

METMEN IN WARTIME

A MILITARIZED WEATHER SERVICE?

Before the agreement for the British Commonwealth Air Training Plan was negotiated, the Royal Canadian Air Force assumed that airmen could be trained as meteorologists and that, apart from the daily provision of weather observations, no assistance would be required from the civilian Meteorological Division. To be meteorologists the RCAF recruited a number of high school graduates in September 1939 and gave them a six-month course in meteorology thinking that this would enable them to perform all relevant meteorological duties, including forecasting. By the end of the course, however, the Meteorological Division had convinced the RCAF that these men could not successfully do the required forecasting and lecturing and that only university graduates trained in meteorology were capable of doing the work. Accordingly, in March 1940, the Department of National Defence requested Transport to provide two dozen meteorological instructors for duties at RCAF's British Commonwealth Air Training Plan stations.

At that time, there is no record of the RCAF proposing that these men seconded from the Department of Transport to air force bases be put in uniform. But, for the next two to three years, rumours continued to persist that the Meteorological Division metmen serving with the RCAF would soon be made uniformed members of that force. Many qualified men were attracted by the possibility, applied to the Meteorological Division for employment, and were accepted. Then, after completing the course and posting to an RCAF station, they became aware of the large pay differential between them and such uniformed non-flying training officers as the navigation instructors. Also, the metmen paid for their room and board in the Officers' Mess that was, of course, free to uniformed officers.

To the metman, a uniform would mean not only more pay but also the opportunity to share in such benefits allowed uniformed people as reduced train fares and, after the war, veterans' preferences and possible financial benefits. As a result, most metmen were anxious to be in uniform. What was not fully appreciated by the youthful metmen at that time was the freedom they enjoyed on the RCAF stations. Not being in uniform, metmen were not called upon to go on parades, to serve as Duty Officer, or to participate in other military activities.

Although prohibited from leaving their positions in meteorology after the first two years of the war, the physically fit metmen were subject to military call-up for army service under the 1940 National Resources Mobilization Act. But, after a metman made his first appearance before a board for a physical examination, the Meteorological Division provided a written explanation of the essentialness of his work every six months and postponements of call-up became a routine matter.

John Patterson was of the opinion that Canada was not big or wealthy enough to afford more than one meteorological service and that fragmentation of the Meteorological Division was something to be avoided at all costs. This fragmentation had taken place in the United States where there were three major military weather services

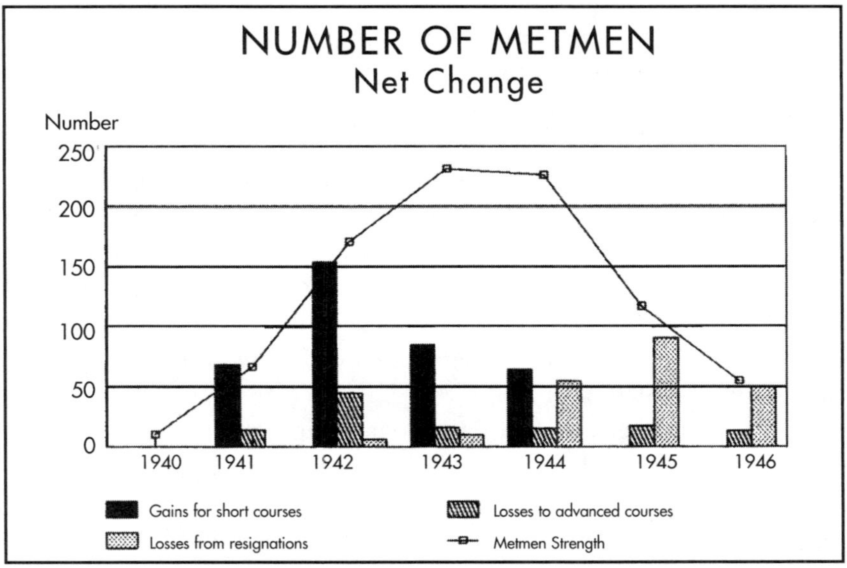

FIGURE 2

The number of metmen on staff in the Meteorological Division each calendar year. Also shown are the annual gains from men entering the short courses and annual losses to the advanced courses and from resignations.

in war years in addition to the civilian Weather Bureau. But, Patterson continued to be convinced that the British method of having the civilian Meteorological Office staff handle all Royal Air Force meteorological requirements in Britain was the right method, and before the war, had managed to convince the RCAF that such an arrangement should be used in Canada.

By 1943, although a supplement of $30 a month had been introduced for metmen serving on RCAF stations, discontent amongst them increased. An association of wartime meteorologists was formed to work towards militarization and strike action was considered. The matter was discussed in such senior government agencies as the Treasury Board and the Civil Service Commission, as well as in the departments of Transport and National Defence, and different solutions were considered.

Finally, in May 1943, John Patterson advised the metmen that the "Air Council have approved in principle the enlistment or commissioning of the Meteorological Staff now working with the RCAF."[23] Problems would remain, however. Militarized metmen could not be returned to civilian status as candidates for the advanced course and promotion to the level of meteorologist. The pay of uniformed metmen would be more than that of their peers who worked in completely civilian offices. Would the salaries of those meteorologists at DAFOs forecasting for the air force be raised to a higher level than the salaries of those forecasting for Trans-Canada Airlines in the same office? These were a few of the problems that would arise by commissioning metmen.

So, within weeks, John Patterson's original views prevailed as Treasury Board ruled that the Meteorological Division must remain intact and continue as a completely civilian organization. By early 1945, though, as the RCAF planned for a Tiger Force of bomber squadrons in Asia to participate in the war against Japan, uniformed meteorological officers were to accompany the force. When volunteers were called for, most metmen responded and the first draft of twenty or so was commissioned in the RCAF in June. More were scheduled to follow early in August but when Japan surrendered and the Tiger Force plan abandoned, the new meteorological officers were discharged and returned to the Meteorological Division.

RELEASE

By the summer of 1944 the need for aircrew was decreasing rapidly as the air war over Europe was well on the way of being won. The British Commonwealth Training Plan had come to an end earlier that year and many schools had been closed although ground services, including meteorology, were maintained at several "shadow stations." This allowed itinerant aircraft to be served and provided a hedge against something happening to the prospects for peace. Further, the RCAF continued to conduct training at a few schools for other forces such as the Royal Navy's Fleet Air Arm.

Many former schoolteachers who had been recruited and trained to be metmen sought and obtained their release to return to their teaching positions in the fall of 1944. In the spring of 1945 all remaining metmen were advised to obtain new positions since it was expected that there would be an excess of meteorological officers.[24] As it happened, the need for metmen decreased at about the same rate as men wished to leave the service and so few, if any, who wanted to remain in meteorology were forced to leave.

At the end of the war in Europe, the total strength of metmen and meteorologists who had first been trained as metmen was about half of what it had been at its peak a year earlier. By early 1946, there were only about fifty-five metmen left of the 373 who had been trained and employed in the Meteorological Division. Most of these fifty-five had served with the air force and had decided to make a career of meteorology. The Meteorological Division had no trouble placing them in the expanded post-war service.

[19] Personal archives, "Duties of Meteorological Officers at Air Training Schools," CIR No. 169, April 8, 1941 and "General Notes for the Guidance of Meteorological Officers," CIR No. 179, May 5, 1941.

[20] AES, Archives, Meteorological Division, "Information for Applicants Regarding the Position of Meteorological Assistant Grade 3," CIR No. 214, October 12, 1942.

[21] AES, Archives, Acting Deputy Minister, Air Services, Department of National Defence to Deputy Minister of Transport, July 1940.

[22] Personal archives, "Relations between Royal Canadian Air Force authorities and

civilian meteorological officers," from Air Force Administrative Order A.29/4 issued March 31st, 1941.

[23] Personal archives, CIR No. 385, May 7th, 1943, signed by J. Patterson.

[24] Personal archives, "Release of Meteorological Officers," Circular No. 790, August 17th, 1945.

CHAPTER 5

AIR DEFENCE 1939–1941

The Royal Canadian Air Force became purely a military organization in 1936 when the Department of Transport was formed. The air force had previously been somewhat involved in civilian aviation activities but now, as Douglas has written "Home defence was the principal justification of the RCAF after 1935."[25]

Two years later, the RCAF became an independent force and began to report directly to the Minister of National Defence. When the three chiefs of staff of the army, navy and air force laid out a Defence of Canada Plan in 1938, the RCAF became responsible for cooperating with the naval forces in seaward defence and the protection of shipping, for cooperating with the army in providing spotting aircraft to detect and calibrate long-range fire by coastal artillery, and for operating fighter squadrons for defence against air attack.[26]

At that time, the RCAF did not have a meteorological capability. It was to aid the air force in carrying out their home defence responsibilities that the Deputy Minister of National Defence wrote to his opposite in Transport in 1938 requesting that the Meteorological Division provide flying forecasts for areas in which the air force was or might become operational and requested that meteorologists be posted to RCAF bases.[27]

METMEN IN WARTIME

THE 1939 RCAF

When the war began, the RCAF's Home War Establishment (HWE) consisted of operational commands on the Atlantic and Pacific coasts — Eastern Air Command (EAC) with headquarters at Halifax and Western Air Command (WAC) with headquarters at Vancouver. A total of ten or so active squadrons had been fairly evenly divided between the coasts but the Munich crisis of September 1938 caused a shift of Canadian focus to the Maritimes. The next year, contracts were written for construction of new airfields at Sydney, Yarmouth, and Debert in Nova Scotia. When war threatened in August 1939 the only permanent squadron on the East Coast was one at Dartmouth because of the lack of airfields. That month other squadrons based in western and central Canada were ordered to fly east to temporary war stations in the Maritimes. At the end of 1939, after some reorganization, there were six operational squadrons in Eastern Air Command whose general roles were either "maritime reconnaissance and strike" or "fighter and army cooperation."

On the Pacific coast, when war broke out, there were two active squadrons left in Western Air Command equipped with but eight serviceable aircraft and two auxiliary squadrons without effective aircraft. By December 1939, there were four squadrons stationed at Vancouver's Sea Island airport and at Patricia Bay airport. There were also some seaplane facilities at Ucluelet, Alliford Bay, and Vancouver's Jericho Beach. Construction had begun or was planned for additional seaplane bases at Prince Rupert, Coal Harbour, and Bella Bella.

At that time, both Eastern and Western Air Commands were under-manned and under-equipped for their home defence responsibilities. There were but 4,153 officers and airmen and only fifty-three aircraft suitable for active operational service. Not one of the units was fully manned and many of the "suitable" aircraft were actually obsolete. Further requests by the Department of National Defence to the government for additional personnel and new aircraft were to receive little or no priority over the first sixteen months of the war because of the more pressing needs elsewhere for new aircraft and trained pilots. There was also significant political pressure from the public and other government departments to send units of the RCAF overseas to the European war theatre. In addition, there was a new and urgent need for pilots to train student pilots in the new British Commonwealth Air Training Plan for which Canada had become largely responsible.

New aero engines and air frames could not be obtained for RCAF operations since, as a result of wartime agreements, the allocation of all munitions and supplies was in British and American hands. Since the danger of an attack on Canada was considered to be remote, the RCAF's home defence establishment ranked very low in the priority lists. Accordingly, their operational program remained small and the demands for aviation meteorological services were not as great as they might have been.[28] Further, it must be noted that, until 1940, the Meteorological Division had no mandate and no resources, either staff or financial, to provide services to the RCAF.

METEOROLOGICAL REQUIREMENTS

The Meteorological Division became aware of the RCAF requirements for operational meteorology in December 1938 when the above-noted letter from the Deputy Minister of National Defence was received. The letter from DND outlined the existing organization of the air force and listed the permanent and temporary bases where squadrons were or might be located.[29]

At that time the RCAF proposed that the meteorological information they were requesting be routinely given to the Eastern, Western, and Air Training Command Headquarters for dissemination by wireless to the stations and units. National Defence questioned whether or not the Meteorological Division was capable of providing forecasts for flying operations on both the Atlantic and Pacific coasts and seaward for distances of 500 miles (800 kilometres). It was stressed that "the most pressing need is the provision of meteorological information, covering the environs of the respective coasts, to the Eastern and Western Air Commands." Within days Transport agreed to cooperate with DND and John Patterson was charged with preparing a preliminary report showing how the Meteorological Division would meet the requirements. Knowing that in the United Kingdom the civilian Meteorological Office provided full meteorological support for the Royal Air Force, Patterson suggested the same be done in Canada. He obtained copies of the pertinent RAF regulations defining meteorological/air force relations and gave copies to both departments. However, negotiations became stalled over the next several months and most contacts between the departments concerned the supply of meteorological observing equipment to the air force stations.

Apparently impatient, the air force attempted to solve the meteorology problem on its own. Even before the outbreak of war, the RCAF had recruited a group of high school graduates for technical training as meteorologists. The RCAF held a training course at the Trenton air station from September 1939 to February 1940 under the misapprehension that the graduates would be able to supply the weather forecasts they required. Little information is available today about the course. However, by the time it was completed, the air force had been convinced by the Meteorological Division that university graduation with mathematics and physics training was a necessary prerequisite for training as an aviation weather forecaster. Another difficulty arose when, during the early months of the war, the RCAF devised and was about to begin an aviation weather observing program in which observations would be taken by airmen using procedures and codes that had not been co-ordinated with the system used on the Canadian and American civil airways.

These matters concerned John Patterson since he had hopes that both the operational and training sectors of the RCAF would agree to use a civilian source of meteorological services as the RAF did in Britain. From other negotiations it appeared to Patterson that the RCAF Air Training Command would accept civilian meteorologists so he saw no reason why the Eastern and Western Air Command operational squadrons could not do likewise. The matter was settled in November 1939 when the RCAF's Chief of the Air Staff agreed to the Meteorological Division's proposal to provide meteorological services for the RCAF and that a conference should be held to work out the particulars.[30]

There was little that the Meteorological Division could do for the RCAF in the early days of the war beyond providing observations from the civilian observing stations and, on the East Coast, aviation forecasts from the newly-established Halifax forecast office. The civilian Meteorological Assistant Grade 2 at Shediac was moved to the Saint

John, New Brunswick, airport in the early war months where his duties were to take the regular weather observations along with pilot balloon and ceiling balloon observations. Other such assistants were posted to Yarmouth and Sydney early in 1940. The RCAF found these civilian "meteorologists" to be quite acceptable and asked for similar men for six additional Eastern Air Command bases and the four Western Air Command bases which would soon be opened. When these could not be provided, the RCAF made the graduates of their 1939–1940 "meteorologists" course non-commissioned officers (NCOs) and posted them to the bases until civilian metmen became available in 1941. At about this time the RCAF dropped the trade name "meteorologist" for these NCOs.

Eastern Air Command also made a strong request for meteorologists, similar to those at the Halifax forecast office, to be posted to each main air base to assist the base controller in making decisions as to whether or not to recall aircraft when weather conditions were deteriorating. It was stated that at Dartmouth, for example, there might be as many as twenty to thirty aircraft on patrol at one time and the local base air controllers needed immediate and continuing forecasts on landing conditions at the base and at a possible alternate base. However, from the RCAF Headquarters in Ottawa, the meteorological liaison officer pointed out that the controllers were not making full and proper use of the meteorological information available to them for planning and monitoring the patrols. It was further noted that the bases were now equipped with teletype and queries could be made to the Halifax forecast office meteorologists at any time.[31]

EASTERN AIR COMMAND

Moving personnel from the Toronto and Montreal District Aviation Forecast Offices, the Meteorological Division had quickly met the RCAF's request for a Halifax forecast office and it was soon able to provide area forecasts and terminal forecasts for the Eastern Air Command bases. The air force wanted similar offices at the operational air bases at Sydney and Yarmouth, Nova Scotia, and at Saint John, New Brunswick, but there were no trained personnel available.

The transmission from Halifax of weather reports and forecasts to the air stations had to be accomplished over the already busy RCAF teletype system because there were as yet no dedicated meteorological circuits in the Maritimes. It is interesting to note that by acting on a good forecast of a severe October 1939 storm in the Maritimes the RCAF was able to save many aircraft by securing them on the ground or by flying them out of danger. This helped significantly in making the RCAF personnel more aware of the importance and value of weather forecasts and information than they had previously been.

Although all air bases required terminal or base reports of existing weather conditions and terminal and area forecasts, the requirements of the different squadrons varied somewhat according to the general roles they carried out. The long-range aircraft used for patrol and convoy escort duties could remain aloft for five or so hours in 1939 and by 1941 the newer models extended the possible time aloft to over six hours. These aircraft had a patrol range of 250 miles (402 kilometres), which was increased to 350 miles (563 kilometres) when new aircraft were acquired.

The patrols were usually out over the ocean to the east on anti-submarine patrol or on convoy escort duty. Other flights took place towards the north over the Gulf of St. Lawrence on anti-submarine patrol. The daily forecasts, issued before dawn each day and valid for the daylight hours for these marine areas, were of great importance to the base controllers and the aircrew. Also of significance in the fog-prone Maritimes were forecasts of the height of the cloud ceiling and the visibility at the bases; forecasts of deteriorating conditions at a base often required the recall of patrolling aircraft or their diversion to land at another base.

Aircraft flying harbour patrol, in cooperative exercises with the army or navy or on fighter sweeps along the coast, were never far from their home bases and so the forecasts were perhaps not as important to them as were the forecasts to those engaged in long duration reconnaissance patrols.

The maritime reconnaissance and strike squadrons, most often equipped with amphibious flying boats or aircraft equipped with pontoons (Supermarine Stranraers, Northrup Deltas, Bolingbrokes, Cansos, Catalinas, and Douglas Digbys) were based at the seaplane facilities and airfields at Dartmouth, Sydney, and Yarmouth. In summer, detachments from these squadrons were moved to seaplane bases at Gaspé or Botwood and to the Gander and St. John's Torbay airports. In addition, a squadron of Lockheed Hudson aircraft flew patrols from Dartmouth during the early war period.

The fighter and army co-operation squadrons based at the Saint John and Halifax civil airfields in the early war months were occupied in patrolling the Halifax harbour and in coastal artillery cooperation work with the army. But theses squadrons were equipped with largely obsolete aircraft; three of them were soon disbanded and one was re-equipped with Grumman Goblins and later Kittyhawks. This squadron flew out of Saint John and Dartmouth airfields in 1940 and again late in 1941. Also, a squadron equipped with modern Hawker Hurricanes was based at Dartmouth and undertook naval cooperation flights before leaving for overseas in June 1940.[32]

The threat to Canada became serious with the fall of France in June 1940 and national defence strategies were changed. Military collaboration began with the United States on the East Coast and joint Canadian-American planning commenced for the defence of all North America. But, efforts by the RCAF in coastal patrolling and convoy escorting continued to suffer from lack of suitable aircraft and pilots as the BCATP training program and support to the overseas squadrons continued to rank higher in the priority lists than the requirements of Eastern Air Command. However, home defence on the Atlantic coast did continue to be considered more important than defence of the West Coast until the Japanese attack on Pearl Harbor in December 1941 changed the situation.

The German U-boat campaign in Canadian waters became intensified in the spring and summer of 1940 and the need for longer-range aircraft became very evident. As a result of the August 1940 Ogdensburg Conference on the mutual defence of North America, the first American troops arrived in Newfoundland in January 1941 and the US Navy became involved in the protection of convoys. Eastern Air Command opened No. 1 Group headquarters in St. John's, Newfoundland, and construction of the nearby Torbay airport was completed. The U-boats became very active off Newfoundland in the fall of 1941 and the marked shortage of aircrew in the Command hampered an effective defence.

The expansion of RCAF activities in 1940 and 1941 increased the pressure on the Meteorological Division for forecasts over wider areas and for longer time periods than had been the case early in the war. Within the Division, John Patterson's chief problems in 1940 and 1941 remained much the same as they had been in 1939 — the shortage of trained staff, particularly of metmen, and the need for additional weather reporting stations.

When it did become possible to post metmen to Eastern Air Command bases in April 1941 a circular outlining the duties of the Meteorological Officers (metmen) at the operational stations was issued. In summary their duties were:

a) to prepare at least two weather maps each day,
b) to interpret the fundamental weather forecasts received from the Halifax forecast office and modify the less important details for local conditions,
c) to supervise the work of the airmen assistants in observing, map plotting, and other office work, and
d) to discuss the weather situation with commanders and flight crews in their hangars.

Also, the metmen were reminded that on an RCAF station they should comply with any requests from the Commanding Officer for meteorological advice and information and consult with him in any matter of doubt or difficulty.[33]

FORECASTS FOR EAC

The Halifax forecast office was staffed by forecasters who held MA degrees in meteorology and were officially classed as meteorologists. A staff of four, in addition to the officer-in-charge, provided all the aviation weather forecasts for EAC squadrons stationed at the various Maritimes air bases. In March 1940, Reg Noble, who had organized the office the previous autumn, was recalled to his earlier position as OIC of the Montreal St. Hubert District Aviation Forecast Office and Rube Hornstein took over as the officer-in-charge. The forecasting establishment was increased to five meteorologists by the end of 1941.

Since the patrol aircraft took off at dawn, the Halifax central operations office of Eastern Air Command and the air bases where the squadrons were stationed required a predawn meteorological situation analysis and forecast for those regions for which EAC was responsible. The predawn forecast was valid from then until dusk while a general operations forecast for the next day was also provided to assist the senior officers in planning the following day's operations.

By August 1940 the Halifax office was issuing terminal forecasts and area forecasts three times a day — at 0500 hours valid from 0500 to 1300 hours, at 1100 hours valid from 1100 to 1900 hours, and at 1700 hours valid from 1700 to 0100 hours the next morning. Synopses and analyses were included with each forecast. In addition, special terminal forecasts were issued for Dartmouth, the hub of the Command's operations, at 0300 and 2230 hours.

The forecast office was located in the same building as the EAC central operations office, rather than at the Dartmouth airport, since this location made the provision of weather maps, consultation, and briefings to the EAC senior officers much simpler than would have been the case from an airport office. Periodically, the forecast office sent the current weather map analyses, forecasts, and outlooks over the RCAF teletype system to the flying bases where airmen technicians assisted the aircrew in interpreting them.

The Halifax office also provided meteorological services to the Royal Canadian and Royal Navies. Their requirements centred on forecasts for the coastal districts around Nova Scotia particularly when convoys were arriving at or departing from Halifax. The navy's chart room at the Halifax dockyards was connected to the main Moncton to Halifax airways meteorological teletype circuit which made all weather information and forecasts available there. In addition, direct or "dedicated" telephone connections allowed for immediate consultation with the forecast office and before convoys left for overseas naval officers visited the office. (See Chapter 16 for a more complete account of the Meteorological Division's relations with the wartime navy.)

NEWFOUNDLAND

As noted earlier, the Meteorological Division had established an office at Botwood in 1937 to provide transatlantic forecasts and other aviation meteorology services to the British and American flying boats during the brief summer flying season. The meteorological office was moved to the new Newfoundland airport at Gander in November 1938 and, the next summer, a meteorologist travelled to Botwood whenever a flying boat was scheduled to arrive or depart. In April 1940, both Imperial Airways and Pan American Airlines announced that they would not be flying through Newfoundland that year and consideration was given to closing the Gander base. However, there was a change of plans and, when transatlantic flights were resumed in July, meteorological services were provided. The story of wartime transatlantic meteorology for aircraft ferrying operations is given in Chapter 7.

In addition to the transatlantic services, the Meteorological Division in Newfoundland provided considerable service for Eastern Air Command. This began in the summer of 1940 when the RCAF expanded their patrols into the oceanic areas off the coast of Labrador and stationed flying boats at Canada Bay, Newfoundland. The RCAF requested forecasts for those areas and, because of the increasingly heavy workload forecasting for Eastern Air Command at Halifax, Rube Hornstein, OIC at that office, suggested that Pat McTaggart-Cowan, OIC at Gander, take over responsibility for the RCAF flying out of Newfoundland bases and this was done. The RCAF needs in Newfoundland were similar to those in Nova Scotia — all available observations, pre-dawn forecasts valid until dusk, and an operations forecast for the next day.

Later, in 1940 and 1941, detachments of three EAC squadrons were posted to Newfoundland bases. Anti-submarine patrol flying from June 1940 to April 1941 at Gander was done by a detachment of No. 10 squadron flying Douglas Digby aircraft. This detachment was replaced by one from No. 116 flying Catalina and then Canso flying boats that flew out of Botwood from July to November 1941. When the Torbay airport, near St. John's, was completed late in 1941, Lockheed Hudson Squadron No. 11 arrived and flew from that base.

The same Gander meteorologists who were responsible for the

transatlantic forecasts provided forecasts for the Eastern Air Command squadrons. Metmen were posted to Botwood in the summer of 1941 and to Torbay later that year to brief the pilots. In addition, when No. 1 Group headquarters was set up at St. John's in July 1941 to control the squadrons operating from the island a meteorologist was posted there.[34] The airport at Goose Bay did not open until 1942 but in November 1941 civilian observers arrived at the base and began transmitting observations every three hours by wireless.

EARLY AMERICAN PARTICIPATION

As a result of the North American Defence Agreement and the 1940 negotiations with Britain, the United States was allowed to establish military bases in both Canada and Newfoundland. Accordingly, in the spring of 1941, a detachment of United States Army Air Force meteorologists and supporting personnel arrived at Gander to set up a forecast centre to serve American itinerant aircraft and locally based squadrons.

The American meteorologists apparently had no knowledge of the Gander forecast office or, in fact, of Canadian meteorological operations in Newfoundland.[35] McTaggart-Cowan invited the Americans to participate in the work of the Canadian office in order to gain experience in local and regional meteorological conditions. Thus, in a few months, when a parallel American office was established on the other side of the airfield, the USAAF had personnel with some Newfoundland experience. Later American meteorological offices were opened at the USAAF Stephenville base and the U.S. Navy base at Argentia.

Following the outbreak of war, contacts with the United States Army Air Force and the U.S. Navy became very difficult for John Patterson. Accustomed to operating independently from the civilian Weather Bureau in the United States, the senior American military meteorologists refused to meet with Patterson or other Canadian civilian meteorologists to plan for possible cooperation when operating in Canada. Patterson had to route all his matters of concern to Francis Reichelderfer, chief of the Weather Bureau, who then dealt with the military meteorological services in the United States. In fact, it took assurance from the British Meteorological Office that, in Canada as in Britain, the civilian service was responsible for all meteorological services to the air force before the American military meteorologists would begin to deal with Patterson.

Finally, in August 1941, a meeting was held in Washington attended by Patterson and McTaggart-Cowan from Canada; C.W.G. Daking, the Civil Officer (Meteorology) in the United Kingdom Liaison Mission in Ottawa; Reichelderfer and Ivan Tannehill of the United States Weather Bureau, and several American military meteorologists. At this conference the necessary personal contacts were established and a foundation laid for future co-operation in expanding the observing network, the transmission of data and forecasts, and weather forecasting.[36] It is interesting to note, however, the American senior military flight commanders insisted on being briefed by their own military meteorologists which necessitated the setting up of American military forecast offices in parallel with Canadian civil offices at bases in Newfoundland and later in western Canada.

OBSERVATIONS

In providing the required services for Eastern Air Command, it was perhaps more difficult for John Patterson to solve the lack of observations and observing stations than to recruit and train metmen. At the outbreak of war, in the area covered by Eastern Air Command — eastern Quebec, the Maritime provinces, and Newfoundland and Labrador — there were about thirty-five synoptic weather reporting stations. Twenty-five of these reported by telegraph and the rest by wireless from islands or isolated coastal locations.

In the prewar synoptic weather reporting system only the Gander forecast office, a one-man station at Shediac, New Brunswick, and another at Saint John had full-time Meteorological Division employees. The Shediac observations had been required at Gander for transatlantic aviation forecasting on the New York, Botwood, and Foynes, Ireland, route but with the outbreak of war the Shediac observer was moved to the Saint John airport. The prime responsibility of the Saint John city observer was to administer the stations in the climate observing network in the Maritimes.

At the other stations, the observers either received small salaries as part-time employees or were employed at a radio or wireless station where meteorological observing was part of their normal duties. Four observations a day (at 0130, 0730, 1330, and 1930 hours) were taken at Gander while at most of the other stations only the traditional 07:30 a.m. and p.m. observations were taken and telegraphed. However, because of the increasing transatlantic aviation needs, noon-time (1:00 p.m.) observations had been commenced at nine additional stations in the late 1930s.

In September 1939 the meteorological teletype system was extended from Montreal to Halifax and a full twenty-four-hours a day observing program was commenced at the Blissville, Saint John and Moncton airports. In Nova Scotia, RCAF airmen observers began similar observing programs at Dartmouth airport, the Sydney seaplane base, and at Yarmouth airport early in 1940. As described in chapter 3, the Meteorological Division managed to get new stations established to take three observations a day at Ecum Secum and Liverpool. The Halifax telegraphic reporting station at the Citadel ceased reporting at the end of 1940 and was replaced by one at the Navy League building, near the forecast office, where Meteorological Division employees took the observations. On Sable Island, the observers began reporting every two hours in September 1941. Except for the observing programs begun by the RCAF at their new Summerside and Pennfield Ridge airports there were no additional full-time observing stations established in the EAC region until 1942.

It proved very difficult to get the needed additional observations each day from the traditional telegraph and wireless reporting stations. At the former, the observers usually had other occupations and were not available for mid-day observations although it did prove possible to add a third observation at three more of those stations in 1940–1941. At the wireless stations, usually manned twenty-four hours a day, the operators had radio monitoring tasks, which kept them at their radio sets, and they were unable to take additional observations.

Before the war there were no observing stations in the interior of the Quebec-Labrador peninsula and it was not possible to establish new stations there until 1942. The stations along the North Shore of the Gulf of St. Lawrence used the existing telegraph facilities but these offices were not open twenty-four hours a day, limiting the number of observations that could be transmitted. Some of the isolated stations around the gulf sent their reports by radio to Fame Point where the observations were put into the meteorological system, but it proved very difficult to increase the traditional meteorological workload of the wireless operators.

WESTERN AIR COMMAND

Although the Department of National Defence recognized that the RCAF could respond more quickly than the army or navy to any hostile threat along the coast of British Columbia, Western Air Command remained under-equipped and under-manned for some months after the outbreak of war. This was not only because of the great need in Eastern Air Command for aircraft and pilots but also because it was assumed that the American fleet would stand between Canada and any attacker from the Pacific Ocean.

In September 1939, RCAF Squadrons No. 6 and No. 4 were based at Vancouver's Jericho Beach seaplane base. The latter squadron, flying Stranraer and Vickers Vancouver aircraft, flew coastal patrols while the former, with Blackburn Sharks, stood by as a striking force. Three auxiliary squadrons, largely equipped with inadequate and obsolescent aircraft, were stationed at the civilian Sea Island airport. It was fortunate that no enemy appeared on the Pacific coast during this quiet period since Western Air Command did not have the aircraft or aircrew to significantly respond.[37]

By May 1940, the building of these new bases had proceeded far enough to allow part of No. 111 squadron, on coastal artillery cooperation duties, and No. 120, a communications squadron, to move to Patricia Bay, No. 4 to Ucluelet and No. 6 to Alliford Bay. This left no operational squadrons at Jericho Beach and only an operational training squadron at Vancouver Sea Island. Further, in September 1940, in an effort to bring East Coast squadrons up to full strength, three of the West Coast squadrons were reduced to a basic minimum in size to the detriment of their capability to increase patrol duties. Also, in January 1941, No. 111 Squadron was disbanded and replaced by a small detachment for cooperation with the coastal artillery.

WAC REQUIREMENTS

When the Royal Canadian Air Force requested meteorological services for Western Air Command from the Meteorological Division in 1940 there were few civilian facilities in Pacific Canada on which to build. The Victoria Gonzales Heights Observatory received observations twice daily by telegraph from about twenty part-time observers in the province and issued public weather forecasts for the southern part of British Columbia. In September 1937, the Vancouver District Aviation Forecast Office began to provide aviation weather forecasts for Trans-Canada Airlines and this required additional observing stations especially along the transcontinental airway. In 1940, the list of new stations included four airport stations where four synoptic and hourly weather observations were reported every day and night.

The forecasters needed, of course, more observations than those from British Columbia. Synoptic observations were received four times a day from stations in the Yukon and the Prairie provinces and from a selection of stations in Alaska and the contiguous United States. Some reports had been received by wireless from ships on the Pacific Ocean but most wireless traffic ceased when the war began. Consequently, the great oceanic expanses to the west of Vancouver Island were largely blank on the Vancouver and Victoria weather maps of 1940. By this time, regular radiosonde observations were taken at some stations in the United States but there were only one or two stations close enough to make their observations of value to the Canadian forecasters.[38]

In November 1939 the Department of National Defence had reached an agreement with the Department of Transport in which the civilian Meteorological Division would provide complete meteorological services to the operational and training commands of the RCAF. At this time the air force was training airmen in a "meteorologist" trade but soon accepted that it would be best if the Meteorological Division would provide trained professionals for forecasting and instruction duties. Some civilian observers were posted to RCAF stations but as the war progressed it became increasingly difficult to recruit qualified men and so practically all observers on Western Air Command stations were airmen by late in the war, although airwomen were trained and posted to some of these stations.

Prior to the spring of 1940 there was little pressure on the Meteorological Division to provide much service for Western Air Command beyond the issuing of terminal and route forecasts by the Vancouver District Aviation Forecast Office. A meteorologist and a technician assistant had been assigned to WAC duties in August 1939 at Vancouver Jericho Beach but, before they could report, they were assigned elsewhere for work with greater priority. Then, in May 1940, when the squadrons were moved to Patricia Bay, Ucluelet, and Alliford Bay, the RCAF requested metmen for each of these stations and for the Coal Harbour, Prince Rupert, and Bella Bella stations, which were to soon open. The metmen were required to interpret the forecasts made at the Vancouver DAFO and to modify them for local conditions, if necessary, and to keep the squadrons advised on the general weather situation. But the Meteorological Division had to advise the RCAF that no metmen were available and so the RCAF posted airmen to each of these bases to take observations and interpret forecasts for aircrew as best they could.[39]

In June 1940 Western Air Command asked for a "forecasting office" at the Patricia Bay RCAF base similar to the one established for the Eastern Air Command at Halifax. Specifically requested were four meteorologists, five meteorological assistants and five teletype operators, all civilians. Also requested were weather reporting stations at the existing and proposed air force bases at Vancouver (Jericho Beach), Ucluelet, Coal Harbour, Alliford Bay, Prince Rupert, and Bella Bella.[40] The Meteorological Division was requested to provide one civilian meteorological assistant for each of the other stations where the RCAF planned to provide three airmen to be trained as weather observers.[41]

The forecast office was wanted immediately and dates were given when accommodation would be available at the other stations. Soon, it was acknowledged that an observing station was not needed at the Jericho Beach base because of the proximity of Sea Island airport and that the forecast office would best be at command headquarters at Belmont House in Victoria. Wireless communications were a problem in these early months in Western Air Command but a new communications system was under development in 1940 and 1941.

As the shortage of trained metmen continued, the Meteorological

Division recruited suitably qualified men and trained them at the Vancouver DAFO to be Meteorological Assistants Grade 2. At the same time the DAFO also trained RCAF airmen to be observers, map plotters, and office assistants. When the necessary accommodation was completed at Patricia Bay in June 1940 a civilian observer and RCAF airmen were posted there. A similar meteorological section was set up at Ucluelet in October but it was late in 1941 before such meteorological sections could be opened at Alliford Bay, Prince Rupert, Bella Bella, and Coal Harbour.

While the Meteorological Division was able to begin staffing action for the required assistants it lacked meteorologists for a forecast office. Accordingly plans were made to link the WAC headquarters by teletype with the Vancouver DAFO and to provide one or two men to interpret the observations and forecasts for the RCAF at their headquarters. Meteorologists did become available in November 1940 and the Western Air Command forecast office became operational at Victoria.

In 1941, after the RCAF had reduced the size of the WAC squadrons to a bare minimum in order to build up the EAC squadrons to strength, there does not appear to have been pressure on the Meteorological Division to significantly increase services to WAC. The WAC forecast office operated with fewer staff than the establishment called for but, by the fall of 1941, communications had improved enough to make it worthwhile to post metmen to the four operations bases. Plans were being made that fall to send new and better aircraft to Western Air Command but, in general, both air force and meteorological needs on the Pacific Coast received low priority until after the Japanese attack on Pearl Harbor in December 1941.

[25] W.A.B. Douglas, *The Creation of a National Air Force — The Official History of the Royal Canadian Air Force*, Volume II (Toronto: University of Toronto Press, 1986), pp. 341–342.

[26] Ibid., p. 373.

[27] AES, File 5930-40-0, Deputy Minister of National Defence to Deputy Minister of Transport, 14 December 1938.

[28] Douglas, *The Creation*, pp. 341–353.

[29] AES, File 5930-40-0, LaFleche to Deputy Minister of Transport, 14 December 1938.

[30] Ibid., G.W. Croil to Chief of Air Services, 22 November 1939.

[31] AES, File 5930-40-1, Noble to Patterson, 24 August 1940.

[32] Information regarding the aircraft used and the areas patrolled has been summarized from *The Creation* (Douglas) and RCAF *Squadron Histories* (Kostenuk and Griffin).

[33] AES, Archives, "Duties of Meteorological Officers at RCAF Stations in Eastern Air Command," Circular No. 171, April 10, 1941.

[34] AES, File 5930-40-0, Noble to Patterson, 14 May 1943.

[35] AES, File 5930-50-2, McTaggart-Cowan to Patterson, 22 May 1941.

[36] Ibid., "Summary of Minutes taken at Informal Joint Meeting," August 8–9, 1941.

[37] Douglas, *The Creation*, pp. 401–402.

[38] Information on early meteorology in British Columbia will be found in Thorne K. Won, *Meteorology in British Columbia; A Centennial Review* (Vancouver: Canadian Meteorological Service, 1971).

[39] AES, File 5930-40-10, "Formation of a Meteorological Service, Western Air Command," 19 August 1940.

[40] Ibid., Slemon to McCauley, 15 July 1940 and earlier correspondence.

[41] Ibid., Organization Order No. 50, 19 August 1940.

CHAPTER 6

EASTERN AIR COMMAND

By 1942, the extent of meteorological services required for Royal Canadian Air Force operations on the East Coast was becoming clearer than it had been during the first year or so of the war. And, the Meteorological Division now had a system in place to train science and technology graduates as metmen and some were available for posting to the meteorological sections at the EAC operational stations. Also, a number of graduates from the short intensive courses for metmen were being given advanced training to qualify them as meteorologists and to prepare them for service at an increasing number of RCAF operational bases as well as at the District Aviation Forecast Offices and the transatlantic forecast offices.

In operational meteorology, the Meteorological Division gave prime attention to the East Coast in the early war years. It was here that the RCAF was most active in carrying out their defence of Canada responsibility by providing coverage for the convoys to and from Britain and in anti-submarine warfare. The RCAF's needs continued to expand on the East Coast in 1942 and there was also an increasing need for transatlantic aviation forecasts to assist in the developing program of ferrying new aircraft over the ocean to Britain. Until December 1941, Eastern Air Command had been given priority over Western Air Command in the procurement of both aircraft and pilots since there had been little or no danger to Canada from the Pacific Ocean. But, in 1942, after the Japanese attack on Pearl Harbor in the previous December, WAC began to receive an increased share of available aircraft and aircrew.

RESPONSIBILITIES

By mid-1941, the German U-boats had begun to hunt within the extreme ocean range of the aircraft then available to Eastern Air Command. But, since the Douglas Digby aircraft based at Gander could only operate within a range of 350 miles (563 kilometres) or so, a number of Consolidated Catalinas were transferred to the Command in early summer and based at Botwood. These aircraft were capable of remaining aloft for 24 hours but, with their slow speed, it took eight and one-half hours to reach a convoy 800 miles (1,287 kilometres) seaward from their base. Then the pilot could devote only four hours to patrolling in order to allow eleven and one-half hours to combat the headwinds on the way home.[42]

In the summer of 1942, the U-boats began to hunt in the Gulf of St. Lawrence and this added a new dimension to EAC's responsibilities. A detachment was sent to the seaplane base at Gaspé to patrol over the gulf and squadrons at Sydney and Dartmouth deployed aircraft into the same area. Further, aircraft and crews in training at schools in the Command were pressed into patrol operations but the operational radius of the available training aircraft was limited to 200 miles (322 kilometres). Important as the gulf was, however, the critical fight was on the ocean routes where Allied shipping losses had reached their highest levels in the winter of 1942–1943.

To help secure the "North American lifeline," Consolidated Liberator aircraft, with an effective patrol range of 700 to 900 miles (1,126 to 1,448 kilometres), were provided to two Newfoundland squadrons. This helped to defeat the submarine "wolf packs" in the fall of 1943 and the anti-submarine war entered a new phase. By August 1944, Eastern Air Command had achieved "an effective coordinated defence of shipping" although the North Atlantic battle against submarines continued for the duration of the war.[43] At the peak of their wartime strength EAC had ninety-four aircraft in seven bomber-reconnaissance squadrons stationed at Maritimes and Newfoundland bases.

In the rest of this chapter attention is paid primarily to the activities of Eastern Air Command and to the efforts of the Meteorological Division in serving EAC aircraft based in the three Maritime provinces. Earlier, when the RCAF began to operate from Newfoundland bases, an arrangement had been made to have the pre-war transatlantic forecast centre at Gander handle forecasts for the squadrons flying from Newfoundland.[44] Chapter 7 deals with meteorological activities in Newfoundland.

SQUADRONS

The number of RCAF squadrons operating in Eastern Air Command increased markedly in the 1942 to 1944 period. Dartmouth continued to be the main hub of activity with three maritime reconnaissance and strike or bomber reconnaissance squadrons operating from that base until the latter half of 1944. Stranraer, Digby, and Hudson aircraft were in use at first but Consolidated Cansos and Liberators were brought in to replace them as soon as possible. Also, two or three bomber reconnaissance squadrons were based at each of Sydney and Yarmouth during this period. To meet the intensified submarine menace in the Gulf of St. Lawrence and around Cape Breton Island, detachments from these squadrons were sent in summer to the Gaspé, Quebec, seaplane base and to the training bases at Chatham, New Brunswick and Mont Joli, Quebec. Also, extensive use was made of the Newfoundland air stations at Gander and St. John's Torbay and later at Goose Bay (after the war the station name was shortened to Goose) in Labrador.

At first, these fighter and army cooperation squadrons, all flying Hawker Hurricanes, were based at Dartmouth and Sydney but, by mid-1942, there was always at least one such squadron at Gander, and from mid-1943 until mid-1944, one and sometimes more such squadrons were based at Goose Bay. There were several Canadian Army cooperation RCAF detachments located at the Nova Scotia bases and, in 1943, Squadron No. 164 was formed as a transport squadron and based at Moncton for the duration of the war.

METEOROLOGICAL REQUIREMENTS

By early 1942, meeting demands from Eastern Air Command for forecasts to cover the coastal and oceanic patrol areas and the operational training schools made for a heavy predawn forecasting load at the Halifax office. Operational squadrons were flying out of Dartmouth, Sydney, and Yarmouth and required forecasts for the day ahead and outlooks for the next day. Similar forecasts were required by the Operational Training Units (OTUs) at Debert, Pennfield Ridge, and Greenwood and at the Charlottetown General Reconnaissance School (GRS) where long range exercises and sometimes operational patrols were flown. All these forecasts were provided by the Halifax forecast office. At each base the RCAF airmen, and later metmen, interpreted the current weather and forecasts sent from Halifax. In addition, the Halifax office forecasted for itinerant RCAF flights out of Dartmouth and other bases in the Command area.

Although the RCAF requested meteorologists from the Meteorological Division for the operational and OTU training bases, it was not possible to provide them because of the short supply prior to the spring of 1942. Pennfield Ridge was the only base to have a meteorologist before that time and this was because of the unusually bad weather on occasion at that base. However, when the first advanced course graduated it became possible to move forecasters with some experience to the operational bases and to replace them at the forecast centres with fresh graduates.

The Halifax aviation forecasts in the early part of 1942 covered Nova Scotia, New Brunswick, Prince Edward Island, the Gaspé peninsula, and the adjacent coastal waters. Fortunately, enough meteorologists became available in May 1942 to open a District Aviation Forecast Office at Moncton and this office took over the forecasting responsibilities for the air training schools in the Command as well as for Trans-Canada Airlines flying into and from Dartmouth. The overload at Halifax was further lessened as squadrons were moved to Newfoundland where their meteorological requirements were provided by the Gander office and, by 1943, by a new Goose Bay forecast centre.

EASTERN AIR COMMAND

EAC FORECAST OFFICE

Note was made in Chapter 3 of the request by the Department of National Defence to the Department of Transport in September 1939 to establish a forecast office to provide service for the RCAF and the Royal Canadian Navy on the East Coast. The organization of this office at Halifax and its activities to 1941 are described in that chapter and in Chapter 5.

Rube Hornstein had succeeded Reg Noble as head of the office in March 1940. The forecasting establishment had been increased to five forecasters in 1940 and later was to be increased to six. The early Halifax meteorologists, like Hornstein, were graduates of the prewar Toronto MA course and all had aviation weather forecasting experience before coming to Halifax. Meteorologists Harvey Halbert, Carl Mushkat, Al Conway, Walter Allen, and Art Grant remained at the Halifax office for some time. Don Ross, Ted Wiacek and such new graduates of the advanced course as William Ganong, Graeme Cameron, and Walter Smith forecasted at Halifax for parts of the 1942 to 1945 period. Some of these forecasters were subsequently posted from Halifax to new offices in the Command where their experience would be of value.

In addition to administering the forecast office as an employee of the Meteorological Division the RCAF expected Hornstein to take on additional responsibilities within Eastern Air Command. The Command senior officers looked to him to monitor, if not supervise, the activities of the metmen and airmen assistants at the various RCAF stations within the Command. As best he could, Hornstein did so on an unofficial basis acting as an advisor to John Patterson in Toronto on one hand and to the RCAF Eastern Air Command in Halifax on the other.

The work at the Halifax forecast office late in the war has been well described by Graeme Cameron, a meteorologist at Halifax in the late war years, writing more than fifty years later:

> Each morning the Chiefs of Staff, Air Vice Marshall Johnson, Admiral Murray and the Army chief officer, with their staff officers would meet in the Operations Room in front of a large wall map of the North Atlantic on which were positioned the various

> convoys. The duty meteorologist opened proceedings with the current weather map covering the North Atlantic and the East Coast. He would then display a second map projecting the weather patterns 24 hours ahead. These maps were displayed side by side until they were overlaid the following morning. . . . We prepared and issued forecasts for the Maritime area from maps drawn up four times every day. We also issued storm warnings for the fishermen and frost warnings for the farmers. Our offices adjoined the Air Traffic Controllers with whom we were in daily contact. We often had sufficient data to draw up a 10,000 foot map. This was helpful in forecasting the direction and rate of movement of weather systems.[45]

The Halifax office also provided specific forecasts and services for the navy and the army which are described later in this book.

After the war, in January 1946, the Halifax office took over responsibility for public weather forecasting for the Maritime provinces, the Gaspé peninsula, and the North Shore of the St. Lawrence. On April 1, 1946, the office became a Dominion Public Weather Office and began a full program of public and marine weather while the Moncton DAFO became responsible for aviation meteorological services in the Maritimes.

METEOROLOGICAL SECTIONS

By 1942 the Canadian meteorological manpower situation had markedly improved although there were still not yet a sufficient number of metmen and meteorologists to fully satisfy the RCAF requirements. Beginning in 1941, the recruiting and training program begun by the Meteorological Division brought twenty or more new metmen into the system every four months. (The lack of instructors and classroom size limited the number of recruits on each course.)

Training in successive intensive short courses continued into 1944; in all, there were twelve such courses that produced 373 new metmen for the Meteorological Division from 1941 to 1944. As described in Chapter 4, the graduates of these courses were officially classed as Meteorological Assistants Grade 3 and were called "the metmen" by the Air Force. The number of available metmen was depleted, however, by the selection of 100 of them, over the 1942 to 1945 period, for the advanced course that prepared them to be meteorologists.

On both the RCAF training and operational stations airmen were provided to the meteorological section for the technical work of weather observing and map plotting. Then, beginning in 1942, airwomen from the RCAF Women's Division were trained for this work and they proved to be very capable. During the last two years of the war the "WDs" largely replaced the airmen in the meteorological sections at the training stations and to a lesser degree at the operational stations.

The first metmen had been posted to Dartmouth, Yarmouth, and Sydney in April 1941 and as more graduates of the short courses became available there were multiple postings to these bases. As well there were initial postings to the North Sydney and Botwood seaplane bases and to Torbay, Gander, and Goose Bay airports. Metmen had been posted to the training schools in the Maritime provinces as early as the summer of 1940 and the number on each station was increased to a maximum of four as more men became available. By the spring of 1944, when the RCAF reached the peak of wartime training and operational activities, there were about twenty-five metmen located at the EAC air bases, the main forecast offices, and the training bases in the Atlantic provinces.

As noted in the paragraphs about the Halifax Forecast Office, newly trained forecasters from the advanced courses were posted there from time to time beginning in 1942. Experienced forecasters for the RCAF Operational Training Units and operational stations did not become available until that year when single postings were made to Debert, Pennfield Ridge, and Greenwood. As more metmen became available, three were posted to each of these Operational Training Units to assist the meteorologist and to brief the aircrews.

Additional metmen were posted to Dartmouth, Sydney and Yarmouth operational bases in 1942, and by 1944 a meteorologist had arrived at each of these bases where each meteorologist was assisted by three metmen. The seasonal Gaspé office was usually staffed with metmen but in the 1943 season, when Group Headquarters No. 5 was located there to co-ordinate the gulf patrols, a meteorologist was put in charge. The fighter OTU at Bagotville and the transport squadron at Moncton were served by metmen. The training schools at Summerside and Mont Joli were staffed with metmen and the office at Summerside had metmen with a meteorologist in charge. Charlottetown was a Royal Air Force "transferred school" and had Air Ministry meteorologists from Britain.

By 1945, staffing with meteorologists and metmen was no longer a problem at the various meteorological sections and offices. There was great uncertainty after VE Day regarding the continuation of meteorological services but the RCAF requested that full service be provided at all stations until further notice. However, the stations were closed fairly rapidly and perhaps two-thirds of the wartime metmen had been released by the fall of 1945. While the exodus of metmen did not cause any crises for the Meteorological Division, the departure of RCAF airmen and airwomen from the meteorological sections did. The RCAF attempted to meet individual requests from airmen and airwomen for discharge and this threatened the observing program at many stations. Consequently, both the Radio and Meteorological Divisions were forced to hire technicians and staff some airport stations many months sooner than had been originally planned in order to maintain the weather observing program.

Activities in the meteorological sections on the various Eastern Air Command bases are given in Chapter 8.

[42] Douglas, *The Creation*, pp. 476–480.

[43] Ibid., p. 597.

[44] AES File 5930-40-1, McTaggart-Cowan to Hornstein, 30 May 1940 and Hornstein to McTaggart-Cowan, 8 June 1940.

[45] Personal archives, a draft manuscript by Graeme Cameron, November 1996.

CHAPTER 7

FERRY COMMAND

As related in earlier chapters the Meteorological Division had established a transatlantic forecast office at the Botwood seaplane base on the north coast of Newfoundland before the war. This office provided services for the experimental British and American flying boat transatlantic flights, which began in the summer of 1937. Anticipating the use of land planes for transatlantic flying, an airport was built at Gander and the forecast office moved there late in 1938. In 1940 the British government decided to use the airport as the main staging base for the ferrying of long-range aircraft to Britain, and North Atlantic aviation forecasts for Ferry Command operations became vital.

CIVIL TRANSATLANTIC FLIGHTS

Each wartime summer the British Overseas Airways Corporation (BOAC), the successor to Imperial Airways, continued flying boat or "clipper" flights through Botwood between North American bases and Foynes in Ireland. These flights required meteorological services in Newfoundland and at Montreal St. Hubert, and later Dorval, airports. During the summers of 1939, 1940, and 1941, a meteorologist was sent from Gander to Botwood with weather maps and forecasts to brief the aircrew each time a flight arrived and departed. Service was also provided to the American clippers that used Botwood in summer; they used a southern Atlantic route in winter.

With increased flying in the 1942 season by the British and American clippers and by an RCAF Canso squadron stationed at Botwood, two metmen were added to the existing staff of one civilian and four airmen assistants. The metmen did the briefing using forecasts prepared in Gander. By the middle of the 1943 season, the workload became so heavy that a meteorologist and a third metman had to be posted to Botwood. The Montreal transatlantic office now supplied advisory forecasts for the eastbound BOAC flights and these were amended at Gander or Botwood as necessary. Pan American Airways and American Express Lines were given complete Atlantic forecasts when they left New York Laguardia airport, but it was the Canadian forecasts issued at Botwood that the aircrew used on the Botwood to Foynes section. Gander and Botwood provided the forecasts for the westbound flights from Foynes to Baltimore and New York.

Because more transatlantic flying was forecast for the 1944 season, an attempt was made to establish a strong independent forecast centre at Botwood but the RCAF was unable to supply the necessary supporting staff. Further, the American clippers were now under the control of the USAAF dispatchers at Gander and so the Botwood staff could only give them information on request. Flying activities decreased, especially at Botwood, when RCAF patrols ceased at the end of the war in Europe but the meteorological office remained open twenty-four hours a day to serve the clipper flights.

In 1941, some months after the ferrying began of land-based aircraft from Gander, the ferry organization contracted with BOAC to operate a Return Ferry Service from Prestwick to Montreal. These westbound Liberator flights brought back aircrew, pilots, navigators and wireless operators, so that they could deliver more new aircraft to Britain. Important military and civilian officials, who needed to travel between Britain and the United States and Canada, were also carried on both the eastbound and westbound flights. The Montreal Dorval transatlantic office was the chief provider of forecasts for the eastbound flights while the aircrew briefing was done at Gander and Goose Bay airports. Over 2,000 Return Ferry Service crossings were made by September 1945 when the service was discontinued.

In 1943 the government asked Trans-Canada Airlines to begin transatlantic flights as the Canadian Government Transatlantic Air Service. The inaugural flight was made in July with a Lancaster between Montreal and Prestwick carrying passengers and mail. (Consideration had been given to have TCA participate with BOAC in the Return Ferry Service but this did not work out.) TCA obtained additional Lancasters and in 1944 three round trips a week were made carrying a million pounds of mail and 2,000 passengers between Canada and the United Kingdom. These flights received similar full forecast services as provided to RCAF operations and Ferry Command. The CGTAS mail and passenger service continued until April 1947 when TCA began overseas flights with North Stars and TCA markings.[46]

ROYAL CANADIAN AIR FORCE

Gander airport, originally called Newfoundland Airport, was largely unused before Squadron No. 10, flying Douglas Digbys, arrived in mid-summer 1940. However, the Digbys' extreme range was but 500 miles which was too limited for effective patrolling and convoy escorting. To rectify this limitation when U-boats began to appear off the Newfoundland coast in the spring of 1941, Allied authorities diverted nine Catalinas to Eastern Air Command from an RAF purchase order. Assigned to Squadron No. 116 these aircraft were based at the Botwood seaplane facility. Cansos were later added to the squadron and both aircraft types were flown from Botwood in the summer seasons of 1941, 1942, and 1943 and detachments were flown from Gander and Goose Bay in 1943–1944.

In July 1941, the RCAF organized No. 1 Group headquarters at St. John's, Newfoundland. This headquarters took operational control of all RCAF units in Newfoundland and was particularly responsible for the squadrons and detachments engaged in air operations in support of the Royal Canadian Navy's Newfoundland Escort Force. The Escort Force had been formed to complete the Allies' system of continuous naval escort of transatlantic convoys between North American ports and Britain. Later in 1941, after negotiations with the Americans, it was agreed that the RCAF would be responsible for patrolling the oceanic area north of 48 degrees latitude to the extreme range of their aircraft off Newfoundland and in the Canadian Coastal Zone west of 55 degrees west longitude.[47]

For short periods of time Catalinas and then Cansos of Squadron Nos. 5, 116, and 162 were also based at Gander. In 1943, No. 10 squadron was re-equipped with Liberator aircraft and continued to fly from Gander. From mid-1942 until mid-1944, one of Hurricane Squadrons Nos. 126, 127, or 129, was based at Gander for home defence.

The Torbay airport opened early in 1942 and was operated by the RCAF. The first squadron based there was No. 11 flying Lockheed Hudsons. No. 145, a squadron flying Hudsons and then Venturas, replaced it. This squadron was also there for most of 1943, and again in 1945. Catalinas and Cansos from No. 5 squadron were at Torbay in 1943 and 1944 and Liberators of No. 11 from late 1943 until the end of the war.

The Goose Bay airport was opened in the spring of 1942 and was also operated by the RCAF. Catalinas and then Cansos from one of Squadrons No. 5, 116, or 162 were based there from June 1943 until mid-1944. For a year in 1943–44 a squadron of Hurricanes (No. 129 and then 130) was located at Goose Bay for home defence.

REQUIREMENTS

The RCAF required full meteorological services at each of the bases — Gander, Botwood, Torbay, and Goose Bay. Because of transatlantic demands, Gander and Goose Bay were already forecast centres with meteorologists for forecasting and metmen to assist in briefing aircrew. The metmen at Botwood used forecasts from Gander each summer until meteorologists were posted there in response to increased activity in the summer of 1944. The Torbay station had a meteorologist in charge and metmen did the briefing. The RCAF requested a forecast office at No. 1 Group headquarters and a meteorologist and metmen arrived in 1943.

The system employed by No. 1 Group in controlling aircraft movement in the Newfoundland area required meteorological offices at the bases as well as at group headquarters. Group headquarters advised the air base commanders at Gander, Botwood, Goose Bay, and Torbay of the operations assignments and these were passed to the squadrons. The Group meteorologist prepared a forecast and issued it to the bases. The meteorologist and/or metman at each base considered the weather and the forecasts for the designated area and advised the local operations officer whether it would be suitable or not. This officer, in consultation with the squadron commander, could postpone or cancel the operation. However, Group could, despite the hazards involved, override the base decision and order the patrol to take place. In any case, once an aircraft was airborne, it came under the full control of Flying Control at No. 1 Group headquarters where responsibility lay regarding recall or diversion of the aircraft to another base if necessary. It was for this reason that Eastern Air Command was very insistent that a strong forecasting staff be assigned to No. 1 Group headquarters.

However, at first, this system did cause much concern to the meteorological staff as well as to the air operations staff and especially the pilots who felt less strain on long flights if they knew that their own base officers were continually watching the weather and would recall them if necessary. The meteorological staff at Group did not deal directly with the patrols and there was confusion when sometimes their forecasts differed from those issued at the bases. This problem was solved by having the meteorologist at Group discuss his map analysis with the base meteorologists in daily telephone conferences. When a consensus was reached, the controllers in the field, the controllers at Group headquarters, and the squadrons all received similar weather advice.

FERRY COMMAND

By the summer of 1940, the British Air Ministry had placed orders in the United States for more than 26,000 aircraft to be delivered at a rate of one thousand aircraft a month. Considering the demand for the limited cargo space available, the amount of time taken to dismantle, crate, load, unload, unpack and reassemble the aircraft, and the increasing U-boat menace, sea transport of the new aircraft to Britain became no longer a secure and efficient method of delivery.

Although a senior RCAF officer had earlier suggested that new aircraft could be flown to Britain using Gander as a staging base, the Air Ministry and the RAF believed the losses from ferrying due to the weather and the lack of navigation aids and emergency fields would be too great. However, when ferrying was proposed to Lord Beaverbrook, then the British minister of aircraft production, he accepted the idea at once and ordered his ministry to work with Canadian Pacific Railway officials to establish and organize an operating company that became the CPR Air Services Department.

Within weeks, and without any formal agreement between the British and Canadian governments, the Department of Transport provided space at St. Hubert airport for a ferrying command headquarters and made facilities at Gander, including the transatlantic weather forecast office, available for ferrying operations. John Patterson, Controller of the Meteorological Division, was unaware of the plans for ferrying aircraft and the need for a weather service until shortly before the first flight was ready to depart.

The first flight, which consisted of seven Hudson aircraft led by Captain D.C.T. Bennett, an ex-BOAC pilot and navigator, took place on the night of November 10–11, 1940, after delays caused by heavy snow at Gander and a subsequent ice storm. All arrived safely at Aldergrove near Belfast, Northern Ireland, in daylight on November 11 after about eleven hours in the air. Later, Bennett praised McTaggart-Cowan for the role he had played in the pioneer flight: "He certainly had an excellent grip on Atlantic weather at that relatively early stage of Atlantic aviation. In fact, he gave us forecasts for the first crossing with practically no ocean information at his disposal; and the result, although not perfect, was extremely valuable."[48]

After several group and individual flights, the first fatal accident happened in February 1941 when a Hudson, carrying a celebrity passenger, crashed near Gander shortly after takeoff with three fatalities. The passenger, Major Frederick Banting, co-discoverer of insulin and Canada's first Nobel Prize winner, was going overseas to help coordinate medical research in Canada and the United Kingdom.[49] Weather had held up the flight at Gander for three days and during this time Banting renewed acquaintances with meteorologist Hugh Bindon, previously one of Banting's assistants at the Connaught Laboratories in Toronto. Later, Bindon stated that he was the last person to speak with Banting on the ground.[50]

There were more lost aircraft and fatalities in 1941 but, with the fall of France and increasing German submarine attacks on Atlantic shipping, the ferrying of aircraft became even more important than expected. To improve the operation several changes in organization were made over the next several months. In May, the agreement with CPR was cancelled and the British Ministry of Aircraft Production took direct control of what became known as the Atlantic Ferry Organization (ATFERO). In September, the new airport at Dorval was opened and ferry operations headquarters moved there. In January 1942 the Royal Air Force Ferry Command (RAFFC) was organized and took over all ferrying operations and a new main transatlantic forecast office was organized at Dorval airport.

Since there was a very short supply of air force pilots and other aircrew when the ferrying was commenced, the aircrew was recruited from civilian ranks. Rewards were high for pilots and radio operators who could each make two trips a month on average. Captains received $500 a trip with expenses, copilots $400 and radio operators $300. Attracted by the good pay most pilots came from the United States while the radio operators were mostly Canadian volunteers on leave from Department of Transport radio stations. By mid-1941 there was an increase in the number of aircraft to be delivered and a shortage of ferrying aircrew. This brought air force personnel into the program as new graduates of the British Commonwealth Air Training Program, pilots, navigators, and wireless operators were formed into aircrews and assigned to the program. The use of these "one-trippers" served another purpose as well since after their delivery flight the aircrew would proceed to RCAF or RAF operational units in Britain.

After the ferry service was militarized other more-experienced RAF and RCAF pilots and navigators were used on short-term assignments to allow them "refresher" training. Both the "one-trippers" and the older, more experienced aircrew underwent ferry training at an Operational Training Unit set up at Debert, Nova Scotia, in 1941 for that purpose. Despite the widespread use of air force personnel it is perhaps surprising that a large percentage of aircrew continued to be civilian until ferrying was discontinued in 1945–46. That this should have continued after the ferrying service was militarized was considered unusual at the time.

The Royal Air Force Ferry Command ceased to exist in March 1943 when the RAF created a Transport Command (RAFTC) and gave it responsibility for all transport operations. Within the RAFTC, No. 145 Group handled transatlantic ferrying operations that continued to be centred at Montreal Dorval airport. (Despite the new organization, most members of the ferry service, other airmen, and metmen continued to use the term Ferry Command as long as the service lasted.)

The new aircraft arrived at Dorval from aircraft factories in the United States and aircrews were assigned to them. Following some additional training they departed for Gander or Goose Bay, which opened in April 1942, to await favourable conditions for beginning their transatlantic flight. Gander and Goose Bay continued to be the main airports of departure for direct flights of large aircraft to Britain.

The United States Army Air Force (USAAF) arrived at Gander in May 1941 and soon the USAAF Air Transport Command began a parallel ferrying operation. In 1942, to move the American Eighth Air Force to Britain, a Northeast Ferry route was developed to allow aircraft smaller than the heavy bombers to fly to Britain by steps from Goose Bay using air bases in Canada's southeast Arctic, Greenland and/or Iceland.[51] In addition to the North Atlantic routes another prime ferry route was a southern one from the Montreal and American airports over the West Indies, through Natal, Brazil, to Ascension Island, West Africa and Egypt to deliver replacement aircraft to the eastern Mediterranean war theatre.

REQUIREMENTS

The capability of the Meteorological Division to respond to the needs for transatlantic aircraft ferrying operations was greatly enhanced by the prior existence of the Gander forecast office. This office and the meteorological organization in Newfoundland had been kept intact to serve the flying boat flights that were resumed in August of 1940 and the RCAF that began operating patrols from Gander that summer. Thus, when the Meteorological Division was asked to provide full meteorological service for aircraft ferrying operations in September, without prior warning, this could be done.

Forecasts for the crossing were issued at Gander until the Dorval office was opened early in 1942. Briefing of crews continued at Gander, and later also at Goose Bay, where forecasts were altered if necessary. These bases were also responsible for supplying additional forecasts and information to the aircraft as they flew eastward and came under control of their United Kingdom destination base.

As described in a chapter on transatlantic meteorological services in *Forecasts for Flying,*[52] the North Atlantic was divided into zones according to longitude and a page of forecast conditions was provided to the aircrew for each zone. Temperatures and winds for various heights up to 20,000 feet (6.1 km), cloud conditions, visibility, and the possibility of ice accretion were forecast for use in both ferrying operations and anti-submarine patrolling. Along with detailed flight plans and weather forecasts, compass headings were provided to the aircrew for each zone to keep the aircraft to the south of any centres of low pressure to provide a self-correcting factor in their flight plans.[53]

It must be noted that much depended on the forecasts and advice provided by the meteorologists. No flight captained by an inexperienced pilot was allowed to depart from Gander or Goose until favourable conditions were predicted along the route. Even experienced pilots could not leave without clearance from the meteorological section. As Christie states "When conditions were marginal the final word rested with Patrick McTaggart-Cowan, whose extensive knowledge of North Atlantic weather earned him the nickname McFog."[54]

Pat McTaggart-Cowan had early gained a tremendous reputation for his Atlantic crossing forecasts. Sir Frederick Bowhill, commander of Ferry Command, wrote in 1942 that McTaggart-Cowan probably did more than any other individual to advance the science and art of Atlantic forecasting. Christie adds in his *Ocean Bridge* "all who reminisce about their Ferry Command experiences have nothing but praise for McTaggart-Cowan."[55]

Since wireless had to be used for the exchange of weather information between Gander and other meteorological centres, this information had to be transmitted in secret code after the war began. It was not until February 1942 that teletype connections between Halifax and St. John's were completed using cable facilities and the teletype was extended to Gander by landline in August. There were still many difficulties but improved technology allowed better facilities for transatlantic exchanges as the war continued.

McTaggart-Cowan's responsibilities as Principal Advisor to the Royal Air Force Ferry Command were by no means limited to the North Atlantic route. He was a member of a joint committee to coordinate procedures for air flight control involving the RAFFC and the United States Air Transport Command. This involved numerous trips of inspection to bases in the United Kingdom, along the planned Northeast Ferry Route, along the South Atlantic ferry route through the West Indies and Brazil to Africa, and an exploratory ferry route from the United Kingdom through Mexico to Australia.

The total number of aircraft ferried is estimated to be about 10,000 but there is no available count of the various types of aircraft ferried across the oceans by Ferry Command. Lockheed Hudsons were the first to be ferried from Gander to the United Kingdom and were probably the most numerous. The record indicates there were thirty-one of them lost, more than any other. Other types with high losses on the North Atlantic route were the made-in-Canada de Haviland Mosquitoes with twenty losses and the North American B-25 Mitchells with 16 losses. There were sixteen Martin A-30 Baltimores lost but these all occurred on the South Atlantic route. The greatest losses in personnel occurred when three Return Ferry Service Consolidated B-24 Liberators crashed in Scotland with a total of fifty-four aircrew and passengers aboard the three aircraft. In all, more than 500 aircrew and about sixty passengers lost their lives on Ferry Command flights.

Some of the aircraft losses were due to the weather, others from equipment failure, and still others from pilot error. Since flying weather forecasts were very important to ferrying operations it is interesting to note that of the thirty-one Hudsons lost, half were lost over the ocean after takeoff from Gander, nearly half from training accidents in Canada, and a few on arrival in the United Kingdom. There were sixteen Mosquitoes lost on the northern route. Of these, seven losses were on the stepping-stone flights over the ocean from

Gander to Greenland to Iceland to the United Kingdom, while six were lost in training in Canada.

That no more than one half of the Ferry Command losses occurred on delivery flights over the ocean speaks well for the meteorological forecasts and the "go or wait" decisions made by both the meteorologists and the departure control officers.

GANDER

By June 1940, after the RCAF had recognized the strategic position of the Gander airport, five Douglas Digby aircraft from Squadron No. 10 had arrived for patrol duty. Except for a brief interval in 1942 this squadron, later equipped with Consolidated Liberators, remained at Gander until mid-1945. Periodically, especially in the winter months, Atlantic reconnaissance squadrons such as Nos. 5, 116, 123, or 162, flying amphibious Catalinas and then Cansos, were also based at Gander airport. One of fighter Squadrons Nos. 126, 127, or 129,

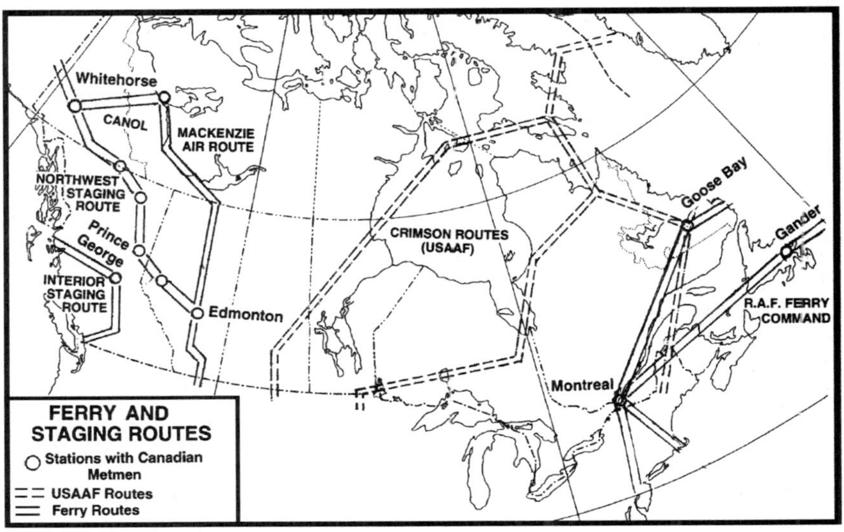

FIGURE 3

Aircraft ferry routes and staging routes in Canada during wartime. The bases where Canadian metmen were stationed are marked with circles.

equipped with Hurricanes for home defence, was on station at all times for more than two years from mid-1942.

The largest user of Gander airport and consequently of meteorological services was the Royal Air Force Ferry Command. The ferrying organization managed at first by Canadian Pacific Air Services, delivered the first Hudsons from Gander to Aldergrove, near Belfast in Northern Ireland, on November 11, 1940. By the end of the war, about 10,000 aircraft had been flown across the Atlantic by Ferry Command, its predecessor and its successor, the Royal Air Force Transport Command.

An increase in the transatlantic exchange of weather information began in late September 1940 which necessitated an increase in the Gander coding staff; there was no coding done at Montreal until the Ferry Command main forecast office was shifted from Gander to Dorval in 1942. Until that time, the St. Hubert office provided weather services for flights from St. Hubert to Gander, while Gander was responsible for the transatlantic flights and those from Gander to St. Hubert.

When the war began, the Gander forecast centre staff consisted of Patrick D. McTaggart-Cowan, officer-in-charge; three other meteorologists; Hugh Bindon; Gib Henry; Keith McLeod; and eleven meteorological assistants. When the ferry flights began and the RCAF increased their patrol flying from Gander, the forecast centre required additional meteorologists. In the fall of 1941 Don Ross joined the staff and over the next four years about twenty different meteorologists forecasted for periods of varying lengths at the Gander office. This number included Dick Douglas, Alex MacVicar, Walter Allen, Ed Barks, Karl Buckthought, Ted Munn, Robie Titus, John Knox, R.W. Glenn, Harold Hutchon, George Brown, Doug Holland, and others who left the service after the war.

In addition to the RCAF patrols operating out of Gander and the great number of ferry flights using the airport as the departure point to Britain, there was an increasing number of transatlantic civilian flights passing through Gander and they also required forecasts and briefings. Under contract with Ferry Command, British Overseas Airways Corporation, using Consolidated Liberator aircraft, began what was called a Return Ferry Service to return ferry pilots, wireless operators, and navigators to Montreal for another delivery flight.

Also, by 1942, American companies, such as Transcontinental and Western Air, began to use Gander on their contract flights to Britain for the USAAF Air Transport Command. Trans-Canada Airlines also gained wartime experience in transatlantic flying by operating what was known as the Canadian Government Atlantic Service using converted made-in-Canada Lancaster aircraft to carry important passengers and overseas mail for the troops in England and later in Europe.

In 1942, when the RAFFC headquarters were moved to the Montreal Dorval airport where a new main transatlantic forecast office was opened, that office became responsible for the basic forecasts. The ferry flights departed from Dorval and at Gander the forecasts were amended if necessary and the crews briefed. Besides the north Atlantic forecasts, forecasts for the southern delivery route through Bermuda, to Brazil and across the ocean to Africa were prepared and issued when required. In addition, when the USAAF began ferrying smaller, shorter range aircraft to Britain using a stepping-stone route from the factories in the United States through Gander or Goose Bay, Baffin Island, Greenland, and Iceland, the Canadian forecast centres prepared the necessary forecasts until the American forecast centres began operating.

With so many demands for service at Gander the forecast centre was divided with one part to serve the RCAF and the other the transatlantic flights. In the last year or so of the war a few meteorologists were allowed to go on transatlantic familiarization flights and remain in Britain for a few weeks on exchange with British forecasters. Until the advent of jet aircraft, more than a decade after the end of the war, when Newfoundland could be overflown, the Gander weather centre remained a very important forecast office in the Canadian meteorological system.

DORVAL TRANSATLANTIC FORECAST OFFICE

The RAF Ferry Command, formed in January 1942, took over all aircraft ferrying work and established its headquarters at Dorval airport. The Montreal District Aviation Forecast Office at St. Hubert airport, established primarily to serve Trans-Canada Airlines, had moved to Dorval airport in September 1941. In the spring of 1942, to serve transatlantic ferrying operations, the Meteorological Division set up a separate forecast office at Dorval to be the main transatlantic forecast office. (This office was commonly called the Ferry Command forecast office.)

Pat McTaggart-Cowan, previously head of the Gander office, moved to take charge of the new Dorval office and became chief meteorological officer for the Royal Air Force Ferry Command and later the Transport Command. Besides the ferry routes through Gander and Goose Bay, the RAFFC headquarters was also responsible for the other more southerly routes to the European and Near East war theatres through Bermuda, the Azores, and Brazil as well as initially for the Northeast or Crimson ferry route through Baffin Island, Iceland and Greenland to Britain. Both the Dorval transatlantic office and the Goose Bay office forecasted for the latter route until the USAAF forecast offices were organized. All briefings of the ferry aircrews for transatlantic flights were done at Gander and Goose Bay.

Staffing of the Dorval transatlantic office began early in 1942, mostly with forecasters who had experience in Newfoundland or at other East Coast offices such as Norm Jefferson, Paul Johns, Clarence Penner, Don Ross, Gib Henry, Ade Lenahan, Hugh Bindon, and Ted Munn. From an establishment of six meteorologists in 1942–43 the staff grew to more than a dozen by the end of 1945. Other meteorologists who continued in meteorology after the war and forecasted in this office in the final years of the war include Don Cameron, Harold Baynton, Robie Titus, Lloyd Richards, Karl Buckthought, Harold Hutchon, Dick Longley, John Miller, Keith McLeod, Bernie Power, and Warren Godson.

Since there was but little or no crew briefing at this office there were no metmen on staff. The office did, however, have a highly regarded staff of female map plotters and assistants recruited from Montreal high school and university graduate ranks. A few years after the war this forecast centre was absorbed by the other forecast centre at Dorval.

GOOSE BAY

The Goose Bay airport was constructed in 1941 and 1942 by the Department of Transport for the RCAF "to facilitate the ferrying of long and medium range aircraft across the Atlantic, to enhance the effectiveness of plans for hemispheric defence, to prevent congestion at the Newfoundland Airport [Gander] and to provide greater security for crews and equipment."[56] Previously, some meteorological observations had been taken near the airport site and in October 1941, after construction had commenced, two observers, a civilian and an RCAF airman, were sent to the site and began sending out observations by radio.

Plans for the use of the airport and the provision of meteorological services were determined at a meeting in February 1942 involving the Royal Air Force Ferry Command and the United States Army Air Force as well as the RCAF and the Department of Transport. In April 1942 Ferry Command began operations at the airport and three meteorologists (Harold Hutchon, Don McClellan and Harold Baynton), along with eight civilian assistants, arrived and the office began providing a twenty-four-hour common service for all requiring it. As usual, however, the USAAF insisted on setting up their own forecast centre to issue forecasts and brief USAAF aircrew.

With the complete absence of landlines all communications to and from Goose Bay had to be done by wireless. In the early days communications with Dorval, headquarters of the RAFFC, were difficult but later both the USAAF and the RAFFC obtained improved wireless communications equipment and a more satisfactory exchange of basic weather data began. In addition to the three meteorologists who arrived in 1942, five others — Larry Campbell, Clarence Penner, Ed Barks, John Knox, and Hank Sabraw — forecasted at Goose Bay in the 1943 to 1945 period. A few other meteorologists who left the service after the war also served at Goose Bay for short periods.

At first, the chief client of the forecast office was the Royal Air Force Transport Command but when RCAF fighter and maritime reconnaissance and strike squadrons were posted to Goose Bay in 1943, a second meteorological office was requested to serve them. But staff could not be provided immediately so a meteorologist was sent to the RCAF office every day to do the briefings. That year, the civilian meteorological establishment at the main Goose Bay office was increased to five meteorologists and two metmen. When the separate RCAF office was opened the RCAF posted four airmen assistants to assist the meteorologists.

With both Canadian and American offices at the airport it was decided that the Canadian office would send out the four synoptic observations and the American office the hourly observations each day. When some difficulties arose over differences in the observations, agreement was reached that the observers must regularly consult and agree on current weather conditions before reporting.

RCAF Hurricane fighter Squadron No. 129, and later No. 130, was stationed at Goose Bay for defence from early 1943 until mid-1944. A detachment from a Canso squadron, first No. 5 and then 162 and finally 116, was stationed there from mid-1943 until mid-1944 when the need for patrols over the North Atlantic Ocean had diminished to the extent that the detachment was withdrawn. Without the Cansos and Hurricanes, meteorological work for the RCAF was limited to issuing routine forecasts for the Goose Bay flying control area and a daily forecast for Squadron No. 164, a transport squadron flying between Goose Bay and Moncton. Consequently, the RCAF meteorological section was closed.

However, work at the main Goose Bay office had increased markedly for the RAFTC and BOAC, and the Trans-Canada Airlines (flying transatlantic flights as the Canadian Government Atlantic Service) frequently used this airport. Although American military activities decreased rapidly after the end of the war in Europe, the Americans wished to retain a military base at Goose Bay and so the Canadian authorities ruled that the RCAF must also remain. This meant that the Meteorological Division had to maintain a full forecast centre at the airport for some time after the war.

AMERICAN COOPERATION

By 1939, the daily Canadian-American exchange of weather data was quite extensive. It included surface and upper air observations, forecasts for inspection and use in offices of the other country, and current information for distribution by radio and teletype to air carriers and others concerned with North American weather. For the first two years of World War II the United States was neutral and many special arrangements had to be made to deny North American and especially Canadian weather information to the enemy. Mention has already been made of the steps taken in Atlantic Canada to accomplish this while providing sufficient information to the American companies for their transatlantic flights.

With the establishment of American bases in Newfoundland in 1941–42, the United States Army Air Force at Stephenville (Harmon Field) and St. John's (Fort Pepperell), and the United States Navy at Argentia, more problems arose. These problems were exaggerated by the reluctance of the American military meteorologists to deal with the Canadian civilian meteorological service although they had been advised that the Canadian service was responsible for all meteorological service, both civil and military, in Canada and Newfoundland, and for transatlantic service west of longitude 30 degree west, by agreement with the United Kingdom and Newfoundland governments.

When a detachment of three forecasters and four observers from the USAAF arrived at Gander in May 1941 to set up a weather office they were apparently oblivious to the fact that there was a Canadian meteorological presence there. But Pat McTaggart-Cowan gave the Americans access to all the Canadian facilities to help them gain experience in dealing with Atlantic weather and they worked in the Canadian office for some time before setting up their own office. The whole matter of Canadian-American cooperation was thoroughly aired at a conference in Washington during August 1941 attended by United Kingdom, United States, and Canadian civil meteorological service representatives and members of the United States Army and Navy weather services. Existing and planned facilities in Canada and Newfoundland were considered and a foundation was laid for cooperation in handling future requirements of the American Armed Forces.

With the entry of the United States into the war in December 1941, the coordination of services began to improve especially after the establishment of a Combined Meteorological Committee, a subcommittee of the Allies' Combined Chiefs of Staff. This committee, consisting of representatives from the United States, the United Kingdom, Canada, Australia, New Zealand, and South Africa became responsible for organizing the necessary facilities and procedures to provide meteorological services for the operation of military aircraft on the various transatlantic ferry routes. This committee met monthly in Washington; John Patterson was urged to have a permanent representative in Washington but declined because of his staff shortage.

Major weather forecast centres were established at Stephenville and Argentia, and at Gander and Goose Bay where they duplicated the Canadian offices. This was necessary because the American military leaders insisted on having their meteorological information delivered to them by an American in uniform. All observations taken by American servicemen at Stephenville and Argentia, and hourly observations from Goose Bay, were distributed on the meteorological communications networks.

To fill gaps in the Canadian observing network, observing stations were established in eastern Canada by the American military services at a number of locations in Newfoundland and Labrador, Quebec and the Northwest Territories. The locations of the stations are given in Chapter 18: Observations.

THE CRIMSON ROUTE

After the United States entered the war it was realized that it would be advantageous and possible to ferry medium-range and fighter aircraft to Britain for the United States Army Air Force and the Royal Air Force. But these aircraft did not have the flying range to allow use of the direct Ferry Command route flown by Hudsons and other long-range aircraft. Other more northerly routes over Greenland and Iceland had to be considered. At least one was required since shipping dismantled aircraft by sea to Britain was too slow and dangerous and shipping losses to the German U-boats were too great when both the air forces desperately needed aircraft for operations in the European theatre.

The proposal for a Northeast Staging Route was put forward by the Americans at a meeting of the Permanent Joint Board on Defence in May 1942, and despite misgivings by the Canadians and the British, the proposal was agreed to. In the Crimson Project, three routes were considered —

a) a northeastern route through Quebec to Goose Bay, and on to bases in Greenland and Iceland,
b) a central route from the Great Lakes region, east of Hudson Bay, to Fort Chimo (Kuujjuad), Frobisher Bay (Iqaluit) and on through Greenland and Iceland, and
c) a western route from California through Regina to Churchill, across the top of Hudson Bay over Southampton and Baffin Islands and then over Greenland and Iceland to Britain.

Many factors had to be considered in the selection of one or more of these routes where extensive and very expensive facilities would have to be constructed. The northeastern route was the shortest and it was developed as part of the main north Atlantic ferrying route used by Ferry Command and is considered earlier in this chapter. Both the central and western routes were considered valuable because most American aircraft factories were in the mid-west and Pacific states and use of either route would be require less time than that required to fly the aircraft to the eastern United States and then north to Greenland and beyond. Weather was another major factor and, in general, better flying conditions could be expected along the central and western routes although to establish these routes as airways would cost much more than the eastern route.

After much consideration and a final meeting involving the Canadian and American military and the relevant Canadian civilian agencies in October 1942, it was decided that the western route would be developed and plans for a central route be dropped. At that time, the Canadian government was of the opinion that neither the western nor the central route was essential and that resources could be better spent elsewhere on the war effort. Consequently, while a new airport was built at The Pas by Canada, the government gave permission to the United States authorities to build airports at Churchill, Coral Harbour on Southampton Island, Fort Chimo (Kuujjuad) in Quebec, and at Frobisher (Iqaluit) on Baffin Island.

In addition to its use as a ferrying route from central and western United States to Britain, the Americans planned to use the northern route for flying an expected large number of casualties from fighting in Europe back to the United States. Since this would be under the direction of the American Red Cross the route was unofficially called the "Crimson Route" and a large hospital was built at Coral Harbour.[57]

Construction of the bases proceeded immediately and USAAF weather forecasting and observing offices were opened at The Pas, Churchill, and Frobisher (Iqaluit). USAAF observing stations on the western route were established at a number of locations in Manitoba, Saskatchewan, and the Northwest Territories. The Meteorological Division continued observing programs at most of these stations after the end of the war (see Chapter 18: Observations). Late in 1945, two metmen, Des Wright and Bill Stewart, were posted to USAAF offices at The Pas and Churchill for a short period to facilitate the transfer of this responsibility.

Although Canadians were not involved in observing and forecasting at the USAAF stations during the war, the communications networks established for the Crimson Route were very useful to the Meteorological Division. Teletype channels were established over landlines from Great Falls, Montana, to Regina and on to Churchill with connections to Edmonton. The Canadian Radio Division station at Churchill gathered the observations from other Canadian subarctic and Arctic observing stations by wireless and used the American teletype circuits to send the data to Regina for relay over the Canadian meteorological circuits. Previously these reports had been sent from Churchill by telegraph to Toronto.

Only experimental aircraft delivery flights were ever made over the Crimson Route. Many USAAF transport, communications, and inspection flights were doubtlessly made over the route. The Royal Air Force Transport Command made a few experimental flights from Edmonton to Britain over the route and some obsolete RCAF operational bombers were flown from Britain to Edmonton for use in training. The USAAF never operated it as a through-ferrying route to Europe although they maintained the stations until September 1945. The route was never used for the return of hospitalized casualties. In the post-war settlements the Canadian government paid the United States government in full for the construction costs and took over the airports.[58]

[46] Canadian aviation officials were anxious to have TCA participate in transatlantic services to facilitate the development of a postwar service. Accounts of this are given in Christie, *Ocean Bridge*, pp. 278–291 and in Philip Smith, *It Seems Like Only Yesterday* [a history of Air Canada] (McClelland and Stewart, 1986) pp. 91–101.

[47] Douglas, *The Creation*, pp. 387–390.

[48] Christie, *Ocean Bridge*, p. 48.

[49] Ibid., pp. 62–72.

[50] AES, Archives, Hugh Bindon interview with David Phillips.

[51] Douglas, *The Creation*, pp. 642–644; T.M. McGrath, *History of Canadian Airports* ([Ottawa: Transport Canada], 1984); and Christie, *Ocean Bridge*.

[52] Morley Thomas, *Forecasts for Flying: Meteorology in Canada 1918–1939* (Toronto: ECW Press, 1996).

[53] Patrick D. McTaggart-Cowan in an interview with David Phillips on October 5, 1983.

[54] Christie, *Ocean Bridge*, p. 142.

[55] Ibid., p. 358.

[56] AES, Archives, Department of External Affairs, "Documents on Relations between Canada and Newfoundland," p. 1389.

[57] Main, *Voyageurs*, p. 178.

[58] An excellent account of the Crimson route is given in Christie's *Ocean Bridge*.

CHAPTER 8

EAST COAST METMEN

Metmen provided most of the weather briefings at the Royal Canadian Air Force East Coast operational stations while the Halifax Eastern Air Command forecast office was staffed with meteorologists. By 1942, when more meteorologists became available, one was posted to most operational stations, to the group headquarters established in the Command, and to the training stations where crews were called out on operational flights from time to time. The meteorologists were trained to issue forecasts for the long distance and long duration flights and also gave briefings but the offices were usually staffed with sufficient metmen to deliver briefings whenever these were required.

On operational and training stations, during World War II, the RCAF most often used the term "meteorological section" rather than "meteorological office," the term used for civilian weather offices. To some extent during wartime, the terms were used interchangeably but in this book an attempt has been made to follow the RCAF terminology. The term "meteorological section" is used when dealing with people, their responsibilities and activities, and "meteorological office" when referring to the physical office occupied by staff in the meteorological section.

Meteorological sections at RCAF stations were created to provide weather information and forecasts as required for flying training and/or operations. At air training schools an additional duty was to present lectures as required for the ground instruction school. The civilian officer-in-charge of the section was directly attached to the officer commanding the station as his advisor in meteorological matters. Also he was directly responsible to the controller (director) of the Meteorological Division on all technical matters.[59]

Space, furniture, telephone, and physical maintenance of the room or rooms occupied by the section, usually in the control tower, were provided by the air force. Equipment used in providing the service, however, such as typewriters and copy machines were the responsibility of the Meteorological Division as were all meteorological observing instruments. The teletype machines were provided by a telegraph company under contract for telecommunication services with the Meteorological Division — Canadian Pacific Telegraphs in Western Canada and Canadian National Telegraphs in Eastern Canada.

The meteorologist, or if one was not stationed there, the senior metman on the station, was named the officer-in-charge of the meteorological section by the Meteorological Division head office. The RCAF airmen and/or airwomen assistants reported to him for all work connected with the section. After the first year or so most meteorological sections were staffed with a sufficient number of airmen and/or airwomen to keep the office opened twenty-four hours a day.

When the operational bases were opened there were not enough metmen available for posting to them so usually one or more RCAF-trained "meteorologists" were posted to each base. These were the twenty or so airmen who had been given a "meteorologist's" course in 1939 before arrangements were made for the Meteorological Division to provide civilian metmen to the RCAF. The untrained airmen who were also posted to the stations were "general duties" airmen and learned weather observing and map plotting on the job. The trained airmen carried a trade classification of "meteorologist" but within months this was changed to "meteorological observer." In time most meteorological assistants qualified for this trade category. Later in the war training courses were organized for airwomen and for some groups of airmen.

When opened, most operational station meteorological sections had an establishment of three airmen but by 1942 most had four or five airmen allowing the section to be kept open twenty-four hours a day, seven days a week. Airwomen began replacing airmen at training stations in the summer of 1942 and while some were posted to operational bases most WDs worked at training stations. Also, by 1942, the senior airman at most stations had been promoted to the rank of corporal and by 1945 some sections had a sergeant and two corporals along with airmen and/or airwomen.[60]

Descriptions of the meteorological sections at the various East Coast bases are given in the paragraphs that follow. Any prior meteorological activities at that location or in the general area are noted. The RCAF squadrons operating from each base are listed with their responsibilities to give an indication of the type of flying the weather forecasts were meant to aid.

The names of meteorologists stationed at each base are given and, when known, the names of the metmen. The posting records for metmen are incomplete since the records of those metmen who left meteorology after the war were not retained.

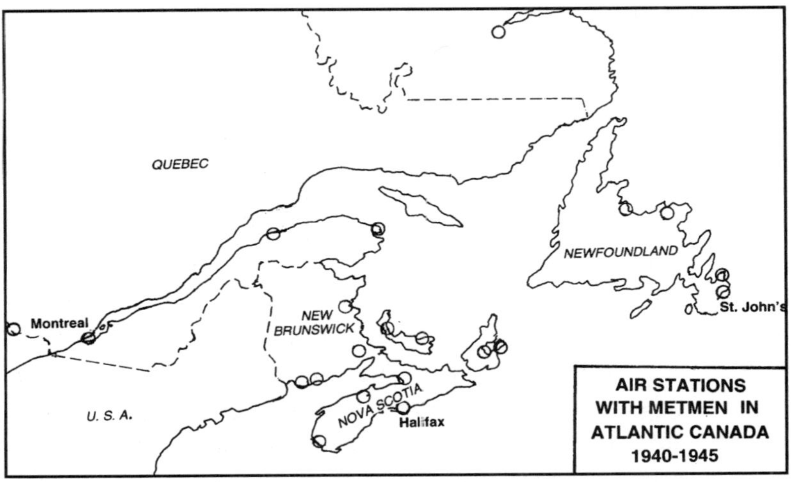

FIGURE 4

Air stations in Atlantic Canada where metmen were stationed in wartime.

DARTMOUTH

During World War I a seaplane base (called Halifax Eastern Passage) was constructed at Dartmouth and was used for a short time by American naval flying boats to protect convoys on the East Coast. Between the wars the base was used by the Canadian air force and in the late 1930s construction began on an airport, which was ready for use by the RCAF at about the time the war began.

Halifax had been a telegraph reporting station for the Meteorological Service since the 1870s but there is no record of observations taken at the Dartmouth air base until the airport was opened. Then, the RCAF commenced a full observing program including hourly observations (sequences), the four regular synoptic observations, and ceiling and pilot balloon observations. The meteorologists at the new Halifax forecast office instructed and supervised the airmen in beginning the observing program.

The Dartmouth air station, which included both the seaplane base and the airport, was Eastern Air Command's operational hub during World War II. From mid-1940 until early 1944 there were three bomber reconnaissance squadrons based there at all times flying on patrol and anti-submarine duty. Squadrons Nos. 5, 10, and 11, equipped with Supermarine Stranraer, Douglas Digby, and Lockheed Hudson aircraft respectively, were there in 1940 and by 1941 these older amphibious aircraft had been replaced by Catalinas and then Cansos. Catalina/Canso squadrons (Nos. 116, 117, and 162) spent the winter months at Dartmouth and when Nos. 5, 10, and 161 were moved to Newfoundland they were replaced at Dartmouth with Squadron No. 111 flying Hudsons and then Liberators, and with No. 145, a Ventura squadron.

From the outbreak of war until mid-1940, fighter Squadron No. 1 flying Hawker Hurricanes was based at Dartmouth before proceeding overseas. From mid-1941 until the end of the war, one of squadrons Nos. 126, 127, or 129 with Hawker Hurricanes was based there for East Coast air defence. An army cooperation flight was at Dartmouth early in the war and from 1942 until late 1945 one or two communications and transport squadrons (Nos. 121 and 167) were also based there. Because Halifax was the administrative East Coast centre for both the air force and the navy there was considerable itinerant air traffic. In 1941, National Defence permitted Trans-Canada Airlines to make Dartmouth airport their Halifax terminal but the Montreal and later the Moncton District Aviation Forecast Office provided the meteorological services for TCA.

At first, the duties of the RCAF meteorological airmen at Dartmouth were taking weather observations and interpreting weather reports and forecasts for the aircrew flying long-range patrols a few hundred kilometres over the ocean to the east and southeast and for other aircrew involved in local coastal flying. Each day a tracing of the most recent weather map and copies of the relevant forecasts prepared at Halifax were delivered to the airport by courier and the airmen were trained to interpret them for the aircrew. Fog was a recurring menace at Dartmouth and the possibility of the airport being fogged in when patrolling aircraft returned was a constant concern to the squadron commanders and those in the meteorological section.

In 1940, the RCAF personnel complement for the Dartmouth meteorological section was a sergeant and six airmen. Metmen had been requested but it was April 1941 before the Meteorological Division could post one to the station. The metman analyzed the 1930 and 0130 hours synoptic maps each day and displayed them in early morning briefing sessions with the pilots. Besides services for the locally based squadrons, forecasts for routes and terminals beyond the area covered by Eastern Air Command were provided to itinerant aircraft.

On account of the increased reconnaissance flying and with patrols going further out over the ocean and operating at all hours of the day and night, a second metman was added to the staff in June 1942 and a third and fourth in 1943 to assist in briefing the aircrews. When the meteorological teletype system was connected to the Dartmouth office in 1942 it was a great improvement over the early system in which the telephone and a courier service had to be used for obtaining meteorological information from the Halifax office. In June 1943, the RCAF began to provide twice-daily weather reconnaissance flights over the coast to observe the extent of fog banks and this information became a great aid to the forecasters and briefers.

The patrols from Dartmouth went increasingly long distances seaward as new long-range aircraft were brought into service. This necessitated a careful watch on the weather to give warning of the possible need to recall the aircraft before the airfield became fog-bound. In February 1944, meteorologist John Dobson was posted to the section as officer-in-charge and the metman establishment decreased to three. In October, the RCAF asked for an additional meteorologist and Walter Smith was posted there. In February 1944, the Dartmouth station became the official Halifax observing station and observations at the city site were discontinued.

In 1945, as patrol flying decreased and the need for briefings decreased, the metmen undertook such extra activities as giving series of lectures for the pilots and beginning a "rawin" program to measure upper winds using radar. After VE Day the patrols ceased and the long-range squadrons left Dartmouth but the meteorological section remained open twenty-four hours a day to serve local RCAF flying. There was uncertainty about the future of the station, but when the RCAF decided to disband their Moncton station, most of the activities formerly carried out there were transferred to Dartmouth and it became necessary to maintain a full-strength office in the immediate postwar period.

Metmen serving at the Dartmouth RCAF operational base in wartime included Burn Lowe, Doug Holland, L.B. Foster, J.A. Burgess, W.W. Judd, Bob Swansburg, and Dave Parkinson.

SYDNEY

Before the war Sydney had long been a telegraph weather reporting station staffed by a part-time observer who telegraphed two observations a day to Toronto. With the war, both the RCAF and the new Halifax forecast office needed more frequent reports from Sydney than the part-time observer could provide and requested the Meteorological Division to post a full-time observer to a North Sydney seaplane base. This was done in November 1939 and airmen observers arrived to take observations under the direction of the civilian observer. Within a few months construction of the Kelly Beach seaplane facilities had proceeded sufficiently to allow the observing site to be moved to that permanent base.

The observing station was moved again in April 1941 when the

Sydney airport construction was completed. For more than a year the RCAF airmen took a full set of observations and interpreted the forecasts for the pilots and squadron commanders. As at Dartmouth, it was April 1941 before the first metman was posted to Sydney where he joined six RCAF airmen and one civilian meteorological assistant. The section depended on the RCAF teletype communications system since a dedicated meteorological circuit was not yet available. The airmen plotted two weather maps each day for analysis by the metman and these were used in aircrew briefings.

No. 8 squadron, with Northrup Deltas and Bolingbrokes for maritime reconnaissance, was stationed at the different Sydney bases until late 1941 when the squadron was moved to Western Air Command. Squadron No. 117, flying Stranraer aircraft and then Catalinas arrived at Kelly Beach and flew anti-submarine patrols over the Gulf in the summers of 1941, 1942, and 1943. Bomber reconnaissance squadrons were based at Sydney airport from early 1942 until shortly after VE Day in 1945. Different squadrons (Nos. 113, 116, 161, and 162) flying Cansos were rotated between Sydney and other bases. One squadron (No. 119) flew Hudsons from the airport in 1942–1943.

Most anti-submarine flying from Sydney was consisted of patrolling the Gulf of St. Lawrence and the coasts of Cape Breton Island. Detachments from the Sydney squadrons were positioned at one or more of Gaspé, Mont Joli, and Chatham during the summer seasons. In 1942 and 1943, either Squadron No. 125 or No. 128 was based at the airport and flew Hurricanes on East Coast air defence. An army cooperation flight was stationed at Sydney from 1941 until 1943.

The meteorological teletype system was extended to Sydney in September 1942 allowing receipt of more weather reports and forecasts than had been possible over the RCAF teletype system. In January 1943, there was a metman and six airmen in the section when meteorologist Larry Cooke was posted to Sydney to handle the increased number of long-range patrols operating out of that base. By July 1943, the section establishment was upgraded to four metmen — three to assist the meteorologist at the airport and one to brief aircrew at the North Sydney seaplane base. That base was closed in December 1943.

In addition to serving the RCAF, the Sydney meteorological section provided forecasts and other weather information to United States Navy Air Transport Service aircraft on their runs between Boston and Argentia, Newfoundland. Also, forecasts were provided to local fishermen, to the Royal Canadian Navy and the Canadian Army, as well as special forecasts for the RCAF Marine Section. After VE Day, although the patrols had ceased operating, the RCAF requested the continuation of a twenty-four-hour service to meet the needs of itinerant aircraft. However, in October 1945, control of the airport passed from the RCAF to the Department of Transport and by the end of the year the metmen were gone and civilian observers had replaced the RCAF airmen.

Metmen serving at Sydney included Lloyd Hillgartner, Keith McGlening, Bob Swansburg, J.A. McCallum, O.K. VanSickle, D.D. O'Reilly, W.F. Doucet, R.B. Ferguson, R.A. O'Brien, and J.A. Ferris.

YARMOUTH

A telegraph reporting station with a part-time observer had existed in the town of Yarmouth for many years before the war and it continued to report until the spring of 1941. In February 1940, a civilian observer (Meteorological Assistant Grade 2) was posted to the newly completed airport and three RCAF airmen soon joined him to begin a full observing program.

Yarmouth was the third major RCAF operational station in Nova Scotia and from early 1942 until VE Day there were usually two reconnaissance squadrons based there. Squadron No. 119, flying Bolingbroke and Delta aircraft, was stationed there in July 1940 but departed early in 1942 and was replaced with Squadron No. 113, with Hudsons and then Venturas, until May 1943. From May 1942 until mid-1945 there was one and sometimes two Canso squadrons (Nos. 162, 160, 5, and 161) flying East Coast anti-submarine duty from this station. A coastal artillery cooperation detachment was stationed at Yarmouth from late 1941 to late 1943. A detachment of a Liberator squadron, No. 10, was briefly on station early in 1942. Most of the anti-submarine patrolling from Yarmouth was done over the Atlantic Ocean out as far as 500 miles (800 kilometres) to the south and east of the base. Weather reconnaissance flights were begun from Yarmouth during the latter part of 1943 and continued until September 1945.

Before a metman arrived in April 1941 to do the aircrew briefing, the RCAF airmen provided information to the crews operating from the station. By this time, sufficient observations were received on the RCAF teletype system from Halifax to plot two weather maps each day for use in interpreting the Halifax forecasts for the pilots. To handle the meteorological work in 1942 it was found necessary to add a meteorologist (H.W. Fleming) to the staff of two metmen, a civilian observer, and three airmen.

In June 1943 a Naval Air Gunners School of the Royal Navy's Fleet Air Arm was formed at Yarmouth and another metman was added to the staff with a fourth in September 1943. Yarmouth had been selected as one of the training centres for heavy bomber squadrons destined for the Tiger Force for action against Japan. A second meteorologist, Ted Wiacek, joined the staff in August 1945 but the training plans were dropped within days after Japan surrendered.

As at other bases, the RCAF then requested that twenty-four-hour service be continued to serve what flying activities were continuing, but early in September the RCAF decided to close the station at the end of that month. The meteorologists and the metmen departed but a small staff of airmen remained to take observations until a Department of Transport radio range staff arrived in April 1946.

Metmen serving at Yarmouth included John Dobson, Robie Titus, J.A. McCallum, Fred Burford, G.V. Patrick, J.E. Tapsell, J.M. Mulholland, and D.N. Huntley.

MONCTON

Radio range operators at the Moncton airport commenced observations in September 1939 when Trans-Canada Airlines began familiarization flights to that airport. In October, meteorologist Hank Sabraw was posted there to furnish Trans-Canada Airlines with the necessary weather information and in January 1940 regular hourly observations were commenced. Later in 1940, when Moncton became a TCA terminal, civilian Meteorological Division observers relieved the radio operators of weather observing and teletype operators arrived.

In December 1940, No. 8 Service Flying Training School was opened at the airport and it trained pilots until January 1944 when it was moved to Saskatchewan. Although no reconnaissance or fighter squadrons were based at the Moncton airport during the war it became a busy RCAF station with numerous transport and itinerant flights. RCAF Squadron No. 164, the air force's premier transport squadron, was stationed there flying supplies from Moncton to Goose Bay, Gander, and Torbay. After Pearl Harbor, the United States Army Air Force stationed a transport squadron (164th) at Moncton for transport between the United States and Goose Bay.

The Moncton District Aviation Forecast Office was opened in May 1942 to meet the increasing needs of Trans-Canada Airlines and the RCAF training schools. This new forecast office provided aviation forecasts for New Brunswick, Prince Edward Island, northeastern Quebec, and the Gulf of St. Lawrence. The Moncton DAFO was completely civilian and not part of Eastern Air Command; it is described elsewhere.

Because the Service Flying Training School meteorological office was physically much closer to the airfield than the DAFO forecast office, the SFTS metmen were regularly asked to provide forecasts and briefings for the itinerant aircraft departing from Moncton. This they were not authorized to do, but the practice was accepted after daily consultations with the DAFO forecasters were made mandatory. When the SFTS was moved from Moncton one metman was left at the airport to continue this service; he was joined by three more metmen by the end of 1944.

The RCAF closed out their Moncton operations in October 1945 and transferred all remaining activities to Dartmouth. The metmen and the RCAF airmen departed by December 1945. Metmen serving in the Moncton RCAF meteorological section after the training school left included Conn Sutherland, Monte Foster, H.D. Skirrow, Frank Upton, Dave Parkinson, and George Washburn.

GASPÉ

When German submarine activity in the Gulf of St. Lawrence became serious in the spring of 1940, a three Stranraer aircraft detachment from No. 5 squadron was posted to the Gaspé seaplane base for patrol work from June to December that year. A forecast office was requested by the RCAF but the lack of trained meteorologists meant that only a civilian observer (Meteorological Assistant Grade 2) was posted there by the Meteorological Division to take observations and give some information to the aircrew.

The only communications landline available was the telegraph line from Campbellton to Gaspé but the local office was not open twenty-four hours a day. Consequently, radio had to be used with the secret code for the transmission of most observations and the reception of forecasts. This proved to be most unsatisfactory since the office was not staffed for coding work. When the flying season ended late in October the station was closed and did not reopen in 1941.

With the return of submarines to the Gulf of St. Lawrence in 1942, a Canso detachment of squadron No. 117 arrived in July. The meteorological section was reopened with two experienced RCAF airmen observers and, in August, a metman arrived. This time an Army teletype system was used and, within a few weeks, meteorological circuit 109 was extended to the base and all the necessary meteorological information became available. The station was closed again in January 1943 after the flying season ended.

Graeme Cameron, a metman at Gaspé in 1942, has written briefly on his work there:

> I was assigned an RCAF corporal and two LACs [Leading Aircraftsmen] who had been trained to take weather observations. . . . We urgently needed a Met teletype . . . we had to work from weather reports squeezed into the Signals Section teletype, along with some EAC regional forecasts. . . . Since the patrols began at daybreak, I would be up each day about an hour earlier to prepare their forecasts. We drew up two daily weather maps based on observations at 0230AST and 1430AST. Between the maps we received hourly reports for the Maritime area. Until our Met teletype arrived, all these reports were received sporadically or not at all.[61]

In the summer season of 1943, Gaspé was made Eastern Air Command's No. 5 Group Headquarters to direct the anti-submarine work over the Gulf of St. Lawrence. The RCAF requested a forecaster and three metmen for the meteorological section and posted in five airmen observers. Meteorologist John Knox arrived and was in charge of the station. Forty years later he told of the situation that summer:

> In 1942 there were two sinking of freighters reported in the Gulf of St. Lawrence, or in the wide reaches of the St. Lawrence River, by German submarines. They [the RCAF] quickly set up a submarine patrol base at Gaspé in response to that U-boat activity. The following spring and summer they decided to increase the patrolling and they created three stations. The headquarters was at Gaspé and the two satellite stations were Chatham, N.B. and Mont Joli, Que. The seaplanes, Cansos and Catalinas, flew out of Gaspé and land-based planes flew from the other two stations. I remember we were in continuous direct communication with Mont Joli and Chatham during this entire operation.[62]

Later in the summer of 1943, with the increasing submarine menace, another metman was sent and meteorological teletype circuit 109 was connected to the meteorological office. Each day two Catalina aircraft made patrols of twelve to sixteen hours. In addition, the navy had a large establishment at the base and requested three forecasts daily. To serve the RCAF and the navy, the meteorological section prepared three forecasts each day for the Gulf of St. Lawrence, using the Halifax forecasts as a guide. Forecasts of winds and visibility were of particular importance. Terminal forecasts for Gaspé, Mont Joli, and Chatham were issued twice daily.

When the 1944 season opened, squadron No. 161, equipped with Cansos, was stationed at Gaspé with detachments at Summerside, Yarmouth, and Seven Islands. Gaspé was not a group headquarters that summer and the meteorological section operated with four metmen. At the close of the season the RCAF observers remained to take observations during the winter season. Squadron No. 5 with Cansos began operations again in May 1945. There were two metmen in the section but with the surrender of Germany activities decreased rapidly and the station was permanently closed in September 1945.

Metmen serving at Gaspé during the war included Graeme

Cameron, G.F. Break, J.A. Ferris, Jack Bocking, C.A. Mutton, and J.C. Jackson.

SEVEN ISLANDS (SEPT-ÎLES)

Clarke City, near Seven Islands, had long been a telegraph reporting station taking the two daytime observations each day. Early in 1942 when the need came for more observations from the St. Lawrence North Shore, arrangements were made to have the nighttime observations taken at Seven Islands and for the local Canadian Government Telegraph station to stay open twenty-four hours a day to transmit the observations. Attempts were made to have existing observing stations to the east of Seven Islands radio their observations to that telegraph station for transmission but this was not successful. Another attempt was made to send observations from Seven Islands to Gaspé via Mont Jacques Cartier on a frequency-modulated radio-teletype circuit but it was only partially successful.

In 1942 construction began on an airport for the RCAF at Seven Islands and a forecast office was requested. By the spring of 1944, when the airport was completed, forecasters were selected for an office and a detachment of Squadron No. 161 arrived for the summer. But, with the decline of the submarine menace, the RCAF activities decreased to the extent that a meteorological section was not warranted. Transport's Civil Aviation Division took over the airport in December 1946.

SAINT JOHN

Squadrons with obsolete aircraft operated from both the Saint John and Halifax civil airfields for the first few months of the war. When construction was completed at Dartmouth and Yarmouth airports, other squadrons were sent there and the original squadrons at Saint John and Halifax were disbanded. An exception was Squadron No. 118, flying Armstrong Whitworth Atlas aircraft, which remained at Saint John until the fall of 1940. The squadron had detachments at Halifax until March 1940 and then at Dartmouth until September of that year. In 1940 an army cooperation flight began flying from the

Saint John airport to test the army's coastal artillery.

A civilian Meteorological Assistant Grade 2 observer was moved from Shediac to the Saint John civil airport shortly after the outbreak of war. By 1941, four RCAF airmen had been posted to Saint John and they carried out a full observing program and briefed aircrew. The civilian observer had departed by the fall of 1944.

DEBERT

The Debert airport was completed in April 1941 and was used immediately by No. 31 Operational Training Unit of the Royal Air Force to train transport aircrew. This station was not part of Eastern Air Command but aircrews in training were from time to time ordered by EAC to assist in patrolling the gulf. Although the RAF/British Meteorological Office sent to Canada a few dozen metmen to staff their training stations this was not the case at Debert and it had to be staffed by Canadian metmen. Graeme Cameron was there early in 1942 and some of his memories of the station are given in the Debert paragraphs in Chapter 14 that deals with the training schools.

The RAF flew Avro Anson and Lockheed Hudson aircraft in their training at Debert. Most of the flying was for long duration and with more meteorologists available, Reid Dexter arrived to head the meteorological section in March 1942. By 1944, a staff of four metmen and several airmen assisted him. In mid-1944 the RCAF took over the Debert OTU and the meteorological section remained as it was until it was disbanded in early 1946.

Metmen who served at Debert during wartime include Graeme Cameron, J.A. Burgess, Doug Holland, S.B. MacKay, H. Sussman, H.W. How, A.L. Wright, Don Black, C.G. Black, O.K. VanSickle, and L. Laking.

MONT JOLI

The Mont Joli airport was constructed in 1941–1942 and was used for both training and operations. In winter, No. 9 Bombing and Gunnery School trained observers, air bombers and air gunners, and in summer, Eastern Air Command used the airport as a coastal patrol base and detachments of several operational squadrons were posted there. Fighter Squadron No. 130 flew from the base during the summer of 1942 along with detachments of reconnaissance squadrons Nos. 11, 113, 119, and 162. Squadron No. 119 was based there in the summer of 1943 and a detachment of No. 5 was there briefly in 1945.

Two and then three metmen were posted to this station to provide forecasts and reports for the detachments and squadrons in the summers. In winters they provided meteorological services to the training school pilots. The school closed early in 1945 and the meteorological section was disbanded when flying operations ceased in mid-1945. The metmen who were stationed there include Keith McGlening, S.E. Deslisle, and L.T. Millar.

SUMMERSIDE

No. 1 General Reconnaissance School, part of the British Commonwealth Air Training Plan system, was located at Summerside from mid-1942 until early 1945. Elmer Stevens, a meteorologist, was stationed here with several metmen for briefing and lecturing to the navigators and other aircrew in training. From time to time crews in training were called out to fly EAC anti-submarine patrols over the Gulf of St. Lawrence.

At the school, the RCAF flew Avro Anson and Lockheed Hudson aircraft. In the summer of 1944, an operational Canso detachment from Squadron No. 161 was stationed there. The meteorological section served both the school and the operational flying. The school closed in February 1945 but the meteorological section remained open until September to serve itinerant RCAF aircraft. Metmen who were there in wartime include Rolly Kendall, Ralph O'Brien, Frank Beard, C.E. Hubley, R.S. Martin, G.K. Higgins, Joe Clodman, O.C. Smith, T.A. Elliott, R.C. Jacka, L.T. Millar, and Harvey Johnston.

CHARLOTTETOWN

The Royal Air Force operated No. 31 General Reconnaissance School at the Charlottetown airport from January 1941 to February 1944. It was a British "transferred school" which became part of the BCATP.

When the RAF left, the RCAF opened No. 2 Air Navigation School at the airport. From time to time training aircraft were ordered to assist EAC by scheduling their exercises over the probable route of U-boats and to carry out other patrols over the Gulf of St. Lawrence in the anti-submarine war.

As long as the RAF was in Charlottetown the meteorological section was manned with personnel from the Meteorological Office/ Royal Air Force. When the RCAF took over early in 1944 six metmen were required. A list of the metmen serving at Charlottetown includes Jim Fowler, A.A. Hoover, O. Tropea, H. Sussman, R.B. Ferguson, L.E. Toombs, D.J. McDonald, C.J. Bishop, C.L. Clendenning, G.V. Patrick, R.M. Skinner, and J.A. McCallum. The school closed in midsummer 1945.

SHELBURNE

The Shelburne seaplane base was used occasionally in winter months by Eastern Air Command's reconnaissance squadrons when more northerly bases were closed because of the danger of ice. Squadron No. 116, flying Catalinas and Cansos, was there for most of the time from November 1942 until June 1943 and No. 117 for two weeks in the early winter of 1943–1944. A meteorological section with metman J.A. Ferris was located there for a short time.

CHATHAM

No. 10 Air Observers School was located at Chatham, New Brunswick from mid-1941 until early 1945. This was a civilian-run school where aircrew were trained and the meteorological office was staffed with metmen and airmen. On occasion, crews in training were called out by Eastern Air Command to assist in patrolling. Also, detachments of operational squadrons were sometimes posted to this station — in September 1942 a five aircraft detachment of Squadron No. 113 flying Hudsons, and from April to December 1943, two Lockheed Hudsons from No. 119.

Primarily at Chatham for the training school metmen in the meteorological section also briefed the operational pilots. The metmen serving there included Reid Dexter, Barney Boville, L.T. Millar, A.P. Loomis, T.A. Elliott, J.A. Doyle, and J.R. Carther.

BOTWOOD

The first transatlantic forecast office had been established at the Botwood seaplane base in 1937. It was closed when the Gander office was opened late in 1938 and during the next two summers a meteorologist travelled to the base whenever transatlantic clipper flights were scheduled.

A detachment of RCAF Squadron No. 116 with Catalinas and later Cansos began flying patrols from the Botwood seaplane base in 1941. When the entire squadron moved there in the summer of 1942 a meteorological section was formed with a staff of two metmen, a civilian observer, and four RCAF airmen assistants. That summer season, and again in 1943, there were as many as nine Canso aircraft flying patrols over the Atlantic Ocean to the northeast of Newfoundland and the squadron required twenty-four-hour forecasts for both the terminal and for the areas to be patrolled.

The meteorological section was located on the second floor of the boathouse with an excellent view of the seaplane landing area. There was no coding staff at Botwood but both telephone and teletype communications with Gander were available to permit the provision of two different aviation meteorological services — one for the RCAF patrols and another for the civil flying boat flights which continued in the summer seasons during the war.

The British Overseas Airways Corporation, Pan American Airways, and American Express Lines operated transatlantic flying boats (sometimes called clippers). The eastbound flights required complete transatlantic forecasts, which were prepared at the Dorval transatlantic forecast office (also called the Ferry Command forecast office) and amended if necessary at Botwood where the crews were briefed. Teletype communications between Gander and Botwood were not good in these early years and frequently the telephone was the only means of communication; wireless could only be used in emergencies since Botwood had no coding staff.

The Botwood meteorological section prepared forecasts for the portion of the westbound flights between Botwood and New York, Baltimore, or Montreal. Botwood also furnished fire hazard forecasts to the Anglo Newfoundland Development Company at Grand Falls. When flying operations ceased at the end of October each year the RCAF squadron was sent elsewhere and the meteorological section activities reduced to taking observations.

When the RCAF planned greatly increased operations for the 1944 season at Botwood, a strong independent forecast centre was required. The Meteorological Division brought in a full map plotting and observing civilian staff and three meteorologists, Ed Barks, John Knox, and Barney Boville, along with two metmen for briefing duties. By this time the American civilian air contractors were under the control of the United States Army Air Force dispatchers at Gander; the Botwood office now had no authority or responsibility for serving the American crews and did so only on request. Further, the USAAF system did not allow any transmission of meteorological information to aircraft in flight and this drastically limited Botwood's possible assistance to the Americans that summer.

In 1945, the meteorological staff sent to Botwood about the first of May included two meteorologists, Boville and Harold Baynton, and two metmen. Because four companies — British Overseas Airways Corporation, Pan American Airways, American Express Airlines, and National Airways — were planning a total of more than ten transatlantic flights a week, the office was open twenty-four hours a day. When the war in Europe ended, the RCAF patrols ceased leaving only the transatlantic operations requiring meteorological services. After the 1945 flying season the meteorological section at the Botwood seaplane base was disbanded.

Metmen stationed at Botwood during the war included Bev Cudbird, Don McCormick, Wilf Hoddinott, and Ned Cutler.

TORBAY

The meteorological office at the newly constructed Torbay airport, near St. John's, opened in the spring of 1942 just after the first aircraft, a detachment of Squadron No. 11 with Lockheed Hudsons, arrived at the airport. From 1942 through 1944 there were always three or four maritime reconnaissance and strike squadrons based at Torbay — No. 145 and then 113 flying Hudsons and later Venturas, No. 5 and then 160 with Cansos, and the No. 11 again in 1943–1944 with Liberators. Fighter squadrons with Hurricanes, Squadron No. 125 followed by No. 128, were based there from mid-1942 to early 1944. In 1942 and 1943 an army co-operation flight was stationed at Torbay and from mid-1943 to the end of the war, a support flight, No. 1 (K), was there.

A meteorologist, Ted Wiacek, was the officer-in-charge throughout the period the meteorological office was open; he had a staff of four metmen and six RCAF airmen (later airwomen). A duplex teletype circuit connected to Halifax was available but connections with Gander had to be by radio or telegraph until the duplex system reached that forecast centre. A telephone connection was available with the American base at Argentia enabling the collection of observations taken at Argentia and Stephenville for transmission by teletype to Halifax.

At first the forecasts used in aircrew briefings at Torbay were prepared at Gander with the local metmen making changes based on local conditions. However, Torbay had its own peculiar orographic features that caused forecasting problems so that in time, and with experience, Wiacek and the metmen prepared their own forecasts for the pilot briefings.

The RCAF patrols required forecasts for twenty-four hours out over the Atlantic to 30 degrees West longitude between latitudes 25 and 52 degrees North. The office at Torbay had added responsibilities as the United States Army Air Force depended on the section for service twenty-four hours a day. This dependency arose since the nearest USAAF weather office was some miles away at Fort Pepperell while the Canadian office was immediately accessible at the airport. Services were also required for Trans-Canada Airlines dispatchers and pilots for their Moncton-Torbay scheduled service that began in 1942. The Royal Canadian Artillery required upper air forecasts four times a day and local St. John's forecasts were provided to the dockyards and construction companies.

At the end of the war, the cessation of the Atlantic patrols and convoy escorting reduced the required work and the airport was taken over by the Department of Transport from the RCAF early in 1946. At that time the metmen and the air force observers departed and civilian observers took over. Metmen serving at Torbay in wartime included Bev Cudbird, Clint Thompson, G.W. Clarke, R.R. Sutherland, Fred Ide, and J.E. Samson.

ST. JOHN'S NO. 1 GROUP

No. 1 Group headquarters was established in July 1941 by the RCAF's Eastern Air Command to control all flying operations in the Newfoundland area. Although the RCAF requested a meteorological section at that time no staff were available, but it turned out that there was no great need for a group headquarters meteorological service for some months. But, by October 1942, meteorological information was needed not only for the RCAF but also for the Royal Canadian Navy, the United States Navy, and the USAAF. Still, it was March 1943 before the Meteorological Division could send a metman to the meteorological section that had been set up with airmen assistants. A few months later, when teletype connections had been made and the RCAF provided a full complement of airmen to assist, a meteorologist (Barney Boville) was posted in along with additional metmen.

The most important duty of the meteorological section at No. 1 Group headquarters was the briefing of the air officer commanding, the chief of operations, and the chief of intelligence at a meeting each mid-afternoon to plan patrols and to issue orders for the next day's operations. The briefing included giving a twenty-four-hour weather outlook for the bases and patrol areas using synoptic and prognostic maps. In addition, strip maps, based on the 0630 GMT observations, were distributed and all operating personnel were briefed daily.

This meteorological section or office was never staffed to be a full forecast centre. The RCAF had requested such a centre but as explained by Meteorological Division headquarters, "Fundamental forecasts for the north Atlantic are prepared at Gander and it is unwise for the meteorological staff at Group headquarters to do much independent forecasting . . . they should interpret the Gander forecasts. . . . St. John's is different from Halifax where there is an experienced and able group of forecasters."[63]

The St. John's personnel were also cautioned not to be dictatorial with the forecasters at the base stations — Botwood and Torbay. Plans for patrols were issued from the headquarters but the weather was a main factor in the decision by the squadrons to fly or not. Once airborne, though, headquarters had full control of a patrol including whether or not the aircraft should return to base when weather conditions began to deteriorate. Consequently it was necessary for consultation between the meteorologists at group headquarters and the metmen at the bases to ensure there was agreement on the synoptic system analysis and forecasts.

Later in the war, when air transport controllers were installed at St. John's, they received full briefings and forecasts several times daily from the metmen. In turn, the controllers, who were in communication with the air transports in flight by two-way radio, supplied the meteorological section with flight conditions between airports including reports on turbulence, icing, height of cloud tops, and so on. Both No. 1 Group headquarters and Air Transport Command ceased operations in the summer of 1945 and the meteorological section was disbanded.

The list of metmen serving at these group headquarters includes Clint Thompson, Don McCormick, Wilf Hoddinott, Bev Cudbird, Fred Ide, A.P.W. Clarke, and R.H.D. Conn.

GANDER

A full account of activities at the Gander forecast office with names of the meteorologists who were on staff there in wartime is given in Chapter 7. Some of the metmen who were stationed there for briefing duties included Bev Cudbird, Bruce Walliser, Don McCormick, and Wilf Hoddinott.

[59] Personal archives, "Regulations for Civilian Meteorological Officers at RCAF Stations," approved by the Royal Canadian Air Force and contained in Air Force Administrative Order A.29/4 issued March 31st, 1941 and issued by J. Patterson as Circular No. 177 on May 2nd, 1941.

[60] Personal archives, letter from Paige Knight (a wartime meteorological observer and after the war a meteorologist), 6 January 1992.

[61] Personal archives, draft manuscript by Graeme Cameron, a wartime metman and later a meteorologist, November 1996.

[62] AES, Archives, John Knox in an oral interview with David Phillips, 31 August 1983.

[63] AES, Archives, File 5930-40-0, Thomson to Noble, 23 May 1943.

CHAPTER 9

WESTERN AIR COMMAND

For the first fifteen months of the war, Western Air Command (WAC) had to deal with obsolescent aircraft, a shortage of aircrew, and a general lack of equipment and facilities as prime attention was given to Eastern Air Command. WAC squadrons made patrols along the coast to track shipping and cooperated with the army but there were no engagements with the enemy. However, during this period facilities were improved, several seaplane bases were constructed, and plans were made for new airports. In 1940, squadrons were moved to the new coastal bases where meteorological sections were established and a new forecast office, dedicated to providing service to Western Air Command, was organized by the Meteorological Division at WAC headquarters in Victoria. But it would be late 1941 before Western Air Command began to grow in strength with more and modern aircraft and new facilities. Fortunately, by that time, the Meteorological Division had trained sufficient personnel to begin to significantly staff the WAC bases.

AFTER PEARL HARBOR

The December 7, 1941, Japanese attack on Pearl Harbor in the Hawaiian Islands completely changed the defence thinking in Canada.

The attack not only brought the United States into the war but the destruction of so many American ships and aircraft increased the operational responsibilities of the Royal Canadian Air Force on the Pacific coast.

With the coast now vulnerable to enemy attack, the public developed fears of a large-scale Japanese assault although senior military officers believed these fears to be unwarranted. Nevertheless, the politics of the situation forced the government to markedly bolster the RCAF's Western Air Command, perhaps more than was necessary, and this caused the RCAF to need more meteorological services.[64]

Prewar RCAF bases had existed at Vancouver's Jericho Beach and at Sea Island, the city's civic airport. Construction of an airport at Patricia Bay, near Victoria on Vancouver Island, was near completion when the war began. By the end of 1941, seaplane bases had opened at Ucluelet and Coal Harbour on Vancouver Island, at Alliford Bay on the Queen Charlotte Islands, at Prince Rupert's Seal Cove, and at Bella Bella on the mainland.

In 1942, Western Air Command began to obtain the aircraft and aircrew it needed to begin to protect not only British Columbia but also to assist the United States in defending Alaska. Canada and the United States had earlier signed a North American defence agreement and the months following Pearl Harbor saw an increasing American presence not only along the Northwest Staging Route in the northeastern part of British Columbia but also along the coastal supply route to Alaska. Soon, an Interior Staging Route, a route from Vancouver through Prince George to Prince Rupert and then by sea to Alaska, was developed, primarily for American transport.

WAC RESPONSIBILITIES

Western Air Command's prime area of responsibility was the Pacific coast and the cities and seaports of British Columbia. Initially, the command was responsible for the entire province but this was reduced to the lower two-thirds of the province in 1943. Command headquarters, at Vancouver until the war, was moved to Victoria. In mid-1940, two bombing and reconnaissance squadrons moved from Vancouver Jericho Beach to Ucluelet and Alliford Bay and another from Vancouver Sea Island to Patricia Bay while the only effective fighter squadron in 1940 was located at Patricia Bay.

In late 1941, reconnaissance squadrons, equipped with Stranraer aircraft, moved to the Prince Rupert Sea Cove, Coal Harbour, and Bella Bella seaplane bases and a year or so later the squadrons were re-equipped with Canso and Catalina aircraft. A Hurricane squadron was at Sea Island in 1942 while one or more fighter squadrons with Hurricanes, Kittyhawks, or Bolingbrokes were at Patricia Bay that year. In April 1942, a fighter squadron was sent to a United States Army Air Force base on Annette Island in Alaska to protect American shipping and the Prince Rupert seaport where supplies for Alaska were transhipped. (At that time there was no Canadian airport at Prince Rupert.) A few weeks later, other squadrons proceeded to Alaska to take part in attacks on the Japanese who had occupied two of the outer Aleutian Islands.

Early in 1942 Western Air Command set up No. 4 Group headquarters at Prince Rupert to control the patrolling activities of the bombing and reconnaissance squadrons at Prince Rupert Seal Cove, Alliford Bay, Bella Bella, and Annette Island. Then, when WAC headquarters moved from Victoria to Vancouver's Jericho Beach in March 1943, Group No. 2 headquarters was established at Victoria to control operations by the squadrons based on the lower mainland and Vancouver Island.

Mention must be made of two other programs undertaken in 1942 which would somewhat lessen the need for RCAF patrol squadrons. After Pearl Harbor, the Americans obtained permission to establish two radar stations on Vancouver Island to be manned by the RCAF. By 1944, about a dozen radar sites had been spotted along the coast from the northern extremity of the Queen Charlotte Islands to the southern part of Vancouver Island. An unsuccessful attempt was made to have one or more of these radar sites provide weather observations. The other program was a scheme for RCAF "coast watchers" to provide an early warning system for Prince Rupert. In 1942–1943 there were "eight detachments of No. 1 Coast Watch Unit" located along the previously uninhabited west coast of the Queen Charlotte Islands. A woodsman, a cook, and two radio operators manned each unit. However, once the radar coverage was in place there was little further need for these stations and the detachments were withdrawn in late 1943.[65]

From late 1942 to early 1944, construction of new airports, such as those at Abbotsford and Boundary Bay near Vancouver and at Tofino, Port Hardy, and Comox, all on Vancouver Island, was completed. These airports were occupied by Western Air Command for operations or for operational training. Another significant advance in those years was the development, by the Departments of National Defence and Transport, of an airway, including airports and communication facilities, in interior British Columbia. This was the Interior Staging Route from Vancouver east and north to Prince George and then west to Prince Rupert. The airports and airstrips were taken over by the RCAF as they were built and Department of Transport civilians manned the radio range stations along the route.

By mid-1944, with the threat of an invasion greatly diminished, Western Air Command began to withdraw squadrons from both the coastal seaplane and airport bases and, by 1945, the only remaining operational squadrons were based at Patricia Bay, Tofino, and Coal Harbour. Transport and communications squadrons continued to be based at Vancouver Sea Island until late in 1945. Except for a Hurricane fighter squadron at Terrace in the winter of 1943–1944, there were no WAC squadrons based at airports on the Interior Staging Route. The United States Army Air Force began to withdraw from the interior route in 1944 and by mid-1945 many facilities were closed.

SERVICES IN 1942 AND 1943

By 1942 the Western Air Command forecast office with a staff of four meteorologists was providing continuous around-the-clock service for the operational squadrons at seaplane bases and airports on the coast. Office support was provided by five civilian meteorological assistants and four teletype operators. A meteorologist was at Patricia Bay with RCAF airmen as observers and map plotters.

Metmen had arrived a few months before 1942 at the RCAF coastal seaplane bases where earlier there had been a civilian observer and three RCAF airmen. A meteorological section was established in 1942 at No. 4 Group headquarters in Prince Rupert where airmen interpreted the observations and forecasts received by wireless from Victoria before metmen arrived.

The teletype facilities which existed in the Vancouver and Victoria regions were not available at the isolated coastal stations where all weather information was sent and received by RCAF wireless. Synoptic weather reports were taken by Transport's operators at several Marine Radio stations along the coast, on Vancouver Island, and the Queen Charlotte Islands. These reports were sent to Victoria and relayed to the other bases. Existing facilities of the Department of Transport on the West Coast were used as much as possible to improve the meteorological system. For example, in December 1942, the air force held meetings with Radio Division officers and with meteorologists in an attempt to obtain more observations from the coastal radio stations and to ask that the Point Grey radio station copy such reports and pass them to WAC headquarters. There they would be transmitted immediately to the RCAF bases rather than having to use the existing time-consuming relay through the Vancouver DAFO. While the District Superintendent of Radio was able to authorize the use of the Point Grey station for relaying observations, he found that his Marine Radio offices did not have enough staff to undertake hourly observations as requested.

In mid-1942 Western Air Command decided to move its headquarters from Victoria to Vancouver and to form a Group headquarters at Victoria to control aircraft from the patrol squadrons. The Command meteorological forecast office accompanied the headquarters when the move took place in March 1943. At that time a new Group headquarters, No. 4 Group, became operational at Victoria and its meteorological section establishment called for a meteorologist, three metmen, and the necessary RCAF plotting staff. A second meteorologist was soon needed and was posted to the Group office.

Each RCAF operational station had one metman on staff by mid-1943 when communications had improved enough to allow the receipt of sufficient observations to warrant plotting maps. Two weather maps were drawn each day by the metmen using map analyses transmitted from Vancouver for their guidance.

George Pincock, officer-in-charge of the Western Air Command forecast office assumed a vital liaison role between John Patterson of the Meteorological Service and the WAC commanding officer. Correspondence of the period shows that WAC requests for meteorological services were sent to RCAF headquarters in Ottawa where the Department of National Defence sent them to the Department of Transport and only then did the requests officially reach the Meteorological Division headquarters in Toronto. This bureaucratic, official channel was slow and time consuming. However, Pincock at Victoria and Reg Noble at RCAF headquarters in Ottawa were usually able to tell Patterson of impending requirements and advise what could and should be done to meet them, well before the official requests were received.[66]

At about the same time the Western Air Command forecast office moved to Vancouver, a new RCAF meteorological wireless circuit was put in operation on the West Coast. The circuit linked seven stations and another seven transmitted their observations to designated stations for insertion on the circuit. In this way hourly weather reports were circulated over the entire twenty-four-hour period. Synoptic observations from the civilian Transport Radio Stations were also transmitted along with a number of selected observations from interior British Columbia, Yukon, Alberta, and the United States stations. Pilot balloon observations were made at Prince Rupert and Coal Harbour and these were transmitted along with relative parts of the radiosonde observations from Tatoosh, Washington State, and Ketchikan, Alaska. All messages had to be coded before transmission and decoded at the receiving end.

The new circuit also carried the forecasts prepared at the WAC forecast office. Synopses and coastal forecasts were prepared four times a day (one based on each of the four synoptic reports), along with forecasts for routes along the coast, a wind aloft forecast, and a future outlook. Also, regional forecasts were prepared each day, just before dawn, and "bombing forecasts," giving expected visibility and weather for the following day were issued daily for Ucluelet, Alliford Bay, and Bella Bella.[67]

Late in the fall of 1943, George Pincock made an inspection trip to a dozen or so of the RCAF stations under the jurisdiction of the Command. The meteorological sections at the coastal seaplane bases – Alliford Bay, Bella Bella, Coal Harbour, and Prince Rupert – were operating twenty-four hours a day and received information by wireless from WAC and other information taken from broadcasts originating in Everett, Washington, and Anchorage, Alaska. Weather maps were prepared twice daily from the 0530 and either the 1130 or 1730 hours synoptic data. At Prince Rupert, observations from the interior of the province were received from Pan American broadcasts at Prince George as well as the observations from Vancouver enabling the meteorologist and a metman to analyze three maps a day. Patrol crews were briefed regularly and maps and analyses supplied to the nearby Group headquarters. The metman at each of the other stations interpreted the WAC office forecasts and briefed the squadron.[68]

It was a different story in the interior where the RCAF was taking over the existing and newly constructed airfields. These developments were chiefly for air transport since there were few, if any, plans to base RCAF operational squadrons at the interior airports. Radio range stations were being built and would be manned by civilian Transport operators who would take the weather observations at most stations although the RCAF planned to have meteorological sections at some bases. Pincock's general remarks on his inspection include the following:

> What impressed me most was the general inadequacy of the meteorological services . . . insufficient information for really satisfactory service [is] being obtained. Teletype installations will rectify this but then we will have insufficient personnel at the various stations to adequately make use of the information being obtained.[69]

SERVICES IN 1944 AND 1945

Inspection of a January 1944 schedule for the Western Air Command Meteorological Wireless Circuit reveals that map analyses and forecasts were coded and transmitted for all four synoptic times. Ciphered observations for map plotting were transmitted for about thirty specific stations. Radiosonde observations from the Canadian station at Coal Harbour and American stations at Ketchikan and Tatoosh were transmitted twice a day. Every three hours the coded weather reports were transmitted for broadcast to pilots.

In 1944 the meteorologists and the metmen at No. 4 Group headquarters meteorological section briefed the flying control officers and the patrol and fighter control officers nine times each day in the headquarters operations room. Weather maps were prepared for the RCAF officers every three hours and vertical cross sections were used in the briefings. When the RCAF made Prince George the headquarters for the Interior Staging route it became necessary to establish a forecast office there. The USAAF already had a forecast office at the airport but disbanded it when the Canadian office became operational in July.

During these months, when the RCAF was rapidly developing new facilities in British Columbia, there was uncertainty regarding their future requirements. Noble, at the RCAF headquarters in Ottawa, was knowledgeable of the RCAF's plans and relayed what information he could to Patterson and they attempted to keep McCauley, OIC of the Vancouver DAFO, and Pincock of the WAC forecast office, in the picture. In February 1944 Noble confidentially advised Patterson that Western Air Command's activities in British Columbia might well be considerably reduced and that No. 2 Group headquarters would likely be disbanded. At about the same time there were indications that the RCAF intended to begin meteorological reconnaissance flights out over the Pacific Ocean.

At this time there were also discussions regarding the necessity of continuing to code meteorological information for radio transmission. The RCAF was now in favour of a partial relaxation but the navy favoured maintaining complete suppression of meteorological information and so this was continued.

Except for Operational Training Units there were few British Commonwealth Air Training Plan schools in British Columbia. There were two Elementary Flying Training Schools in the Vancouver area at Sea Island and Boundary Bay in the 1940 to 1942 period but meteorological sections were not required at these schools since their flying was local. However, the Operational Training Units (OTUs) on the Pacific coast did require meteorological sections. These flying training establishments were part of the BCATP but were also linked with Western Air Command, since on occasion, aircrews in training were ordered out on operational patrols. A Royal Air Force OTU was located at Patricia Bay from October 1941 until June 1944 when the

OTU was moved to Comox and became an RCAF school. Another RCAF OTU was at Patricia Bay from 1942 until the end of the war and a fourth school had training units at both Boundary Bay and Abbotsford from early 1944 until late 1945.

Most Operational Training Units were for training heavy bomber and transport crews and required the scheduling of long flights, both in distance and time. Because of this, by the summer of 1944, the meteorological establishment at each of the OTU bases called for two meteorologists and three metmen, and from the RCAF ranks, one sergeant, two corporals, and five airmen or airwomen. The pilots and navigators had already taken courses in meteorology but the meteorologists at the OTUs gave them some advanced training in the science.

In the fall of 1944 a conference of the officers-in-charge of the WAC forecast office and the four OTU stations was organized. The utilization of staff, office procedures, briefing and debriefing methods were compared, and discussions took place on the adequacy of observations and the communications system as well as of the forecasts and analyses issued from the WAC forecast office. One result of the conference was the scheduling of telephone conferences twice each day at 0500 and 1500 hours. During these teleconferences the WAC forecaster gave his analysis of the synoptic situation and his forecasts. The base meteorologists then commented on the forecasts, in light of the operations planned, and advised of local conditions which might affect the operation. This exchange led to a better understanding at the Command office of the various base weather conditions and how these might affect the planned operation.

Reductions in WAC operational activities had already begun in mid-1944 and continued into 1945. Ironically, soon after the meteorological sections at the OTUs were fully staffed, the training programs began to wind down and all schools and meteorological sections were closed by the end of 1945. In the interior, even before construction and installation of many of the planned facilities were completed, the meteorological system began to disband. Since neither the RCAF or the USAAF had little need for Interior Staging Route forecasts after the spring of 1945, the Prince George forecast office was disbanded and the Vancouver DAFO took over its airways forecasting responsibilities. However, a Prince George dependent office, staffed by three metmen and five civilian assistants, remained open to serve scheduled flights of Canadian Pacific Airlines, itinerant civil and military aircrews, and the public.

OBSERVATIONS

Before the war, the extent of the synoptic observing network in British Columbia, except that part of the province eastward along the transcontinental airway from Vancouver to Alberta, had been very meagre. In 1939 there were but twenty-four synoptic observing stations in the province and seven of these were along the transcontinental airway where most stations took hourly as well synoptic observations. Seven Marine Radio stations along the coast took two observations a day as did less than a dozen other stations, mostly traditional telegraph reporting stations, scattered over the interior of the province.

The need for more observations to improve forecasting for aviation was the driving force behind a remarkable expansion in the size of the network in the Western Air Command region. The story of this expansion is related in Chapter 18: Observations.

WAC FORECAST OFFICE

In August 1940 Western Air Command requested that a full forecasting office be established for the Command. Don McIntyre, a forecaster at the Vancouver DAFO, was sent to Victoria to set up an office at Belmont House, the Command's headquarters. Shortly, George Pincock, also from the Vancouver office, was named officer-in-charge, and with meteorologists McIntyre and Fred Kelly, began operations in November 1940.

In a 1983 oral interview, McIntyre related that the meteorological office was on the top floor of the building and the operations room was below it in the hotel ballroom. Once an hour a meteorologist went down to the operations room to brief the fighter and bomber control officers on present and predicted weather conditions along the coast and out over the Pacific. This was not an easy task since the efforts of the forecasters were hampered by both the lack of weather observations and by wireless communications problems. There were virtually no observations from the Pacific Ocean as only a very few coded ship observations were available. All meteorological information to and from the isolated coastal stations was by wireless; all communications including coding and decoding were the responsibility of the RCAF communications staff.[70]

In 1941, meteorologists Jack Turner and J.A.D. MacNeil were added to the Victoria forecast office staff to assist in issuing forecasts for the patrol squadrons and other RCAF operations at the coastal bases. Also, in 1941, the WAC forecast office took over the public weather forecasting for British Columbia from the Victoria Gonzales Heights Observatory. After Pearl Harbor, the broadcasting of public forecasts for the coastal regions were discontinued but pertinent forecasts were still issued and published for the interior regions.

In March 1943 the forecast office moved with Western Air Command headquarters to Jericho Beach, Vancouver. Meteorologists Ralph Tyner and H.S. Keenlyside arrived at about this time at the WAC office and remained there for most of the war. Ray Walkden, Ken Harry, Bill Gutierrez, and E.N. Ellis also came early in 1943 but soon moved on to other postings. Others who forecasted at the WAC office in 1944 and 1945 were Bill Cameron, Don McMullen, Bill Markham, Joan Griffiths (one of the few women to become meteorologists) and Paul Brun.

Communications were improved with a new meteorological wireless circuit in January 1944 and further improved when the teletype circuits came into operation by the end of March 1944. There had been too few observing stations but with new Transport radio range stations on the coast and new airports and range stations in the interior beginning to provide observations, the information available to the forecasters was greatly increased in 1944 and 1945.

The WAC forecast office continued in operation after the war when it became a Dominion Public Weather Office responsible for British Columbia public weather forecasts. The Vancouver District Aviation Forecast Office took over responsibility for all civil and military aviation forecasting and when space became available at Sea Island airport in 1948 the two offices were combined.

PRINCE GEORGE

A synoptic weather observing station had been established in the town of Prince George in June 1929. At the time the war began the observer took synoptic observations, collected observations by radio from eight fire ranger stations, and gave whatever information he could to pilots landing at the nearby municipal airport. The airport was used by itinerant aircraft and by Pan American Airways whose staff took hourly observations and collected observations from other points by radio. Transmitted from Prince George by Pan American and Yukon Southern radio stations, many of these observations reached the Edmonton weather office and were put on the teletype circuit.

With war in the Pacific, American air traffic to Alaska was greatly increased requiring a metman at the Prince George municipal airport in February 1942. A new airport was then under construction and the metman established an office there and commenced observations in July 1942. For a while observations were taken at three Prince George locations – synoptic reports from the town by the original observer, reports from the new airport by the metman and his staff, and unofficial hourly reports from the old airport by Pan American personnel. The airports were only about seven miles apart but, since their elevations varied by a few hundred feet, local effects often produced variations in the weather and the different "Prince George" weather reports caused confusion to the users.

Communications were a major problem at Prince George since radio had to be used to communicate with the airports until March 1943 when telephone connections were made. At this time radio range operators took over observing at the new airport and three more metmen were posted to provide services for civil transport and transient flights. By the fall of 1943 the USAAF had established a forecast office at the airport with three meteorologists, six observers, and radio operators to serve their transport and aircraft ferrying operations through Prince George. At this time there were three offices providing meteorological information to aviation – the Pan American, the USAAF, and the Canadian offices – but no central office.

In January 1944, when the teletype system was being installed and Prince George became the headquarters for the Interior Staging Route, the RCAF requested the organization of a full Canadian meteorological forecast office. The USAAF agreed to disband their office and have their aircrew use the Canadian office when it became operational. Meteorologist J.R. Henderson arrived early in 1944 and by late spring sufficient information was being received by wireless to allow the plotting and analysis of weather maps. Henderson left for the Vancouver DAFO when four other meteorologists arrived – Don McIntyre (OIC), Don Slater, Ken Harry, and Ralph Tyner who was posted to Vancouver within two months. McIntyre, Slater, and Harry remained at the office until the summer of 1945 and were joined by Sam Neamtan of the Winnipeg DAFO on temporary duty. Metman Gordon Muttit was posted to the Prince George forecast office late in 1944 and was followed after four months by Stu Shannon.

The teletype system had become operational by the end of March 1944 but the organization of the new office was hampered by wartime manpower restrictions which made it difficult for the Meteorological Division to recruit civilian assistants and teletype operators. With the office operational, the American forecast office was withdrawn on July 10, 1944. By late 1944, the RCAF had few flights along the inland route and when American traffic decreased markedly, it was decided to reduce the office to the status of a dependent forecast office and some meteorologists were posted out. But, since the agreement with the USAAF called for a full forecast centre to be available to service American flights, the independent forecast office could not be closed until the Americans agreed. This happened in May 1945 and the remaining meteorologists left that summer and the office reverted to dependent status with Shannon as OIC.

The dependent office in the last few months of 1945 was staffed by three metmen and five assistants. Some Canadian and American military flight traffic continued, there were two Canadian Pacific Airlines scheduled flights each day, and itinerant aircraft arrived and departed. Besides issuing aviation forecasts at the office, public weather forecasts from Vancouver were tailored for the local newspapers and for industrial, forestry, and agriculture interests. The office continued to operate in this manner for some time after the war.

[64] Douglas, *The Creation*, p. 409.

[65] Ibid. pp. 406–408, 422–423.

[66] AES, Archives, File 5930-40-10, numerous letters from Pincock to Patterson beginning in November 1940.

[67] Ibid., 8 June 1943.

[68] Ibid., 3 December 1943.

[69] Ibid., 3 December 1943.

[70] AES, Archives, Donald McIntyre in an interview with David Phillips, 9 November 1983.

The control tower at No. 13 Service Flying Training School, North Battleford, Saskatchewan, in 1944–1945. The meteorological office occupied the second floor of the building.

[COURTESY: METEOROLOGICAL SERVICE OF CANADA]

The control tower staff at No. 10 SFTS, Dauphin, Manitoba, in the summer of 1942. (The civilian is the author, then a metman.)

[COURTESY: MORLEY K. THOMAS, ARCHIVES]

A flight of North American Harvards from a Service Flying Training School. Harvard-trained pilots were posted to squadrons in Fighter Command or remained in Canada as instructors. [COURTESY: NATIONAL AVIATION MUSEUM, ID 27249]

An Avro Anson in flight. Ansons were flown at some SFTSs for twin-engine pilot training and ultimate service in Bomber Command. Ansons were also flown at Air Observer Schools where staff pilots flew the aircraft while student aircrew — observers, navigators, or wireless airgunners — carried out their training exercises. [COURTESY: NAM, ID 23626]

A Cessna Crane. Cranes were aircraft flown by student pilots at twin-engine advanced training schools. Usually these new pilots were then sent to operational training units before posting to squadrons in Bomber Command.
[COURTESY: NATIONAL ARCHIVES OF CANADA, PA-135608]

A Curtiss Kittyhawk on the tarmac. Kittyhawks from Western Air Command squadrons took part in the Alaskan campaign in 1942 and 1943 and patrolled the Pacific coast. [COURTESY: NAM, ID 11015]

[COURTESY: DEPARTMENT OF NATIONAL DEFENCE]

Members of the first RCAF (Women's Division) Meteorological Observer course at Toronto in 1942.

Back row:
W300433 G.H. Jensen; W305064 M.A. Nixon; W304073 A. McFadden; W302967 D.M. Langstaff; W304898 L.M. Algren; W303498 W.E. Rudge; W300388 B.I. McLennan

Centre Row:
W303968 M.J. Pryce; W303596 A.L. Grafton; W304782 E.M. Nedden; W303958 B.E.R. Todd; W303964 P.I. Whitfield; W303818 S.M.F. Playfair; W304904 C.E. Periard; W304190 M.R. Miller

Front Row:
W303803 M.H.A. Potter; W304158 E.M. Cook; W303377 M.E. Elliot; W304781 L.M. Cooper; (W304903 I. Ponder or W301383 A. Penson or reverse); W304071 I. Menu; W304900 D.V.M. Goodman

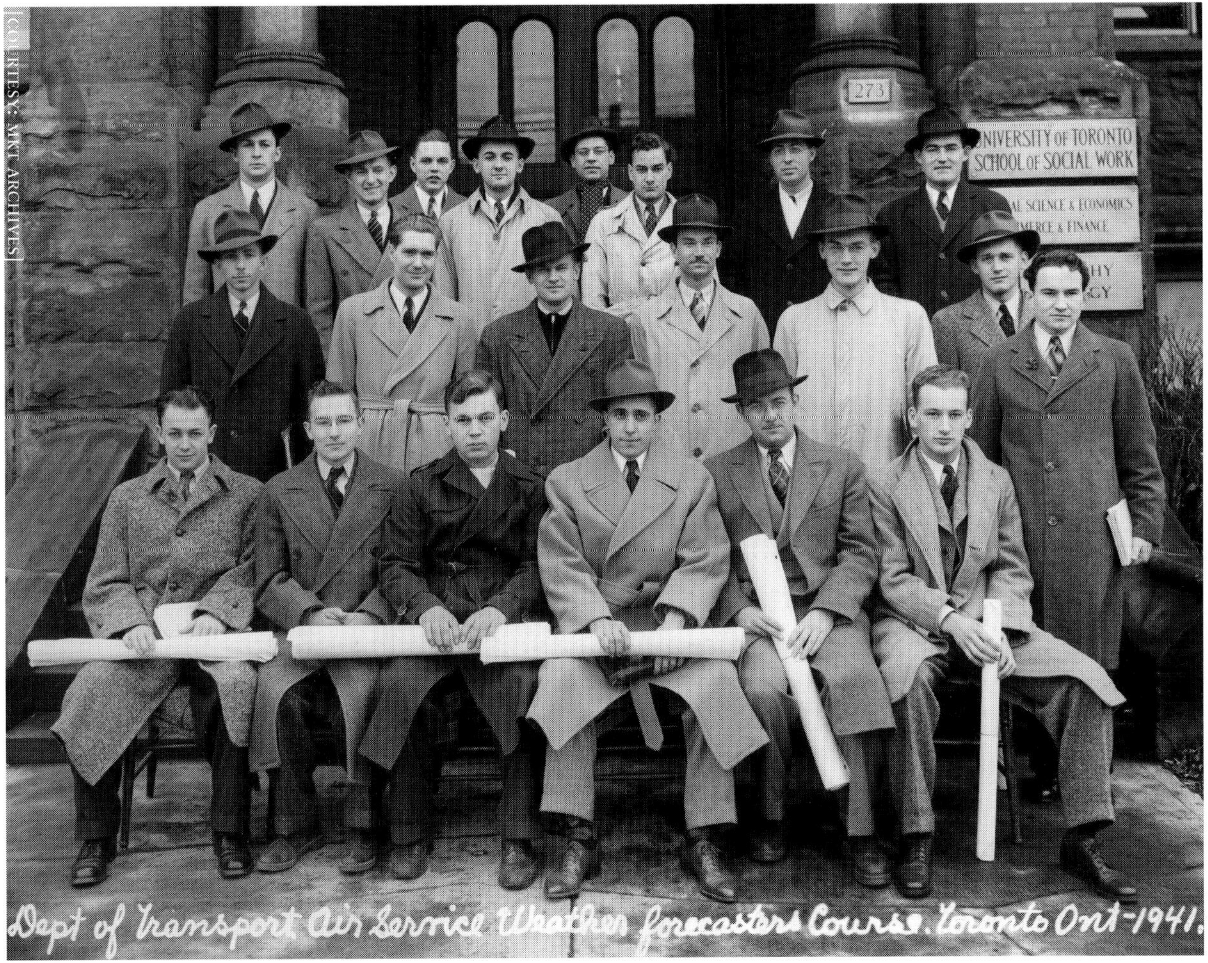

[COURTESY: MKT ARCHIVES]

Members of Fourth Intensive Course in Meteorology at Toronto, Oct. 1941–Jan. 1942.

Front row:
Des Wright, Keith McGlening, Elgin Little, Fred Page, Lloyd Richards, Ray Walkden.

Second row:
Bill Jarmain, Don Day, Walter Halina, Ernie Jones, Ed Barks, Bruce Walliser, Churchill Longman.

Third row:
Morley Thomas, Howard Vaughan, Don Stewart, Graeme Cameron, Don Slater, Bill Cameron.

Fourth row:
Robie Titus, Bev Cudbird.

A Consolidated Canso from No. 162 Squadron ready for a patrol over the North Atlantic in October 1944. Catalina and Canso aircraft replaced Stranraers for coastal patrols in 1941 and 1942. [COURTESY: DND, PL 33838]

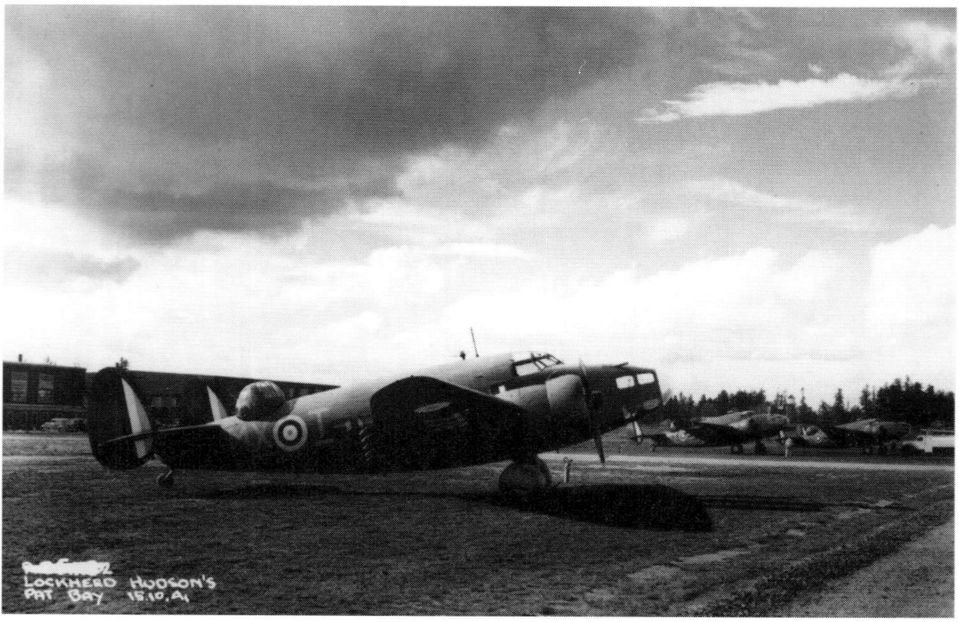

A Lockheed Hudson at Patricia Bay. These aircraft were flown on coastal operations primarily by squadrons in Eastern Air Command. In Naovember 1940 a flight of seven Hudsons were the first aircraft ferried to Britain over the Atlantic Ocean from Gander. [COURTESY: NAC, PC-137369]

A Supermarine Stranraer from No. 5 (BR) Squadron patrolling near Halifax, Nova Scotia, in September 1939. Early in the war most operational squadrons in Eastern and Western Air Commands flew Stranraers on patrol.

[COURTESY: DND, PL 2731]

Three wartime leaders who later became heads of the Meteorological Service: Pat McTaggart-Cowan, Andrew Thomson, and Reg Noble (1965).
[COURTESY: AL MCQUARRIE]

A metman at work at the forecasters' desk, No. 1 Air Observers School, Toronto Malton. [COURTESY: MSC]

CHAPTER 10

WEST COAST METMEN

A general description of wartime meteorological sections at Royal Canadian Air Force stations is given in the introduction to Chapter 8: East Coast Metmen.

PRINCE RUPERT

Marine Radio personnel had taken weather observations at Prince Rupert since 1908. Early in the war a seaplane base was built for the RCAF at nearby Seal Cove and, in September 1941, a Meteorological Division civilian observer was posted there along with three airmen observers. Squadron No. 7, flying Blackburn Sharks and then Stranraers, arrived in December to begin patrol duties. The squadron was re-equipped with Cansos and Catalinas in May 1943 and remained at Seal Cove until April 1944.

In June 1942, No. 4 Group headquarters was set up at Prince Rupert to control the patrol operations of the squadrons at Prince Rupert, Alliford Bay, Bella Bella, and Annette Island. At that time there was no airport at Prince Rupert so RCAF fighter squadrons Nos. 115 and 118 were based at a United States Army Air Force airfield on Annette Island in Alaska, some sixty miles (100 kilometres) or so distant. The squadrons were there from April 1942 to November

1943 to protect the Prince Rupert seaport, an important transhipment port for Alaskan supplies, and the sealane to the Alaskan ports.

The Prince Rupert meteorological section remained at the patrol station on Seal Cove and all meteorological communications were carried over the RCAF wireless network. The civilian and airmen observers provided the squadron, and later the group headquarters, with forecasts received from the WAC forecast office at Victoria. By the fall of 1942 communications had improved to the extent that sufficient observations to plot daily weather maps could be obtained and so a metman was sent there to analyze the maps and brief the aircrew. As patrols were increased, two additional metmen were needed and, in May 1943, a meteorologist, Bill Gutierrez, arrived to head the meteorological section.

In June 1943, Gutierrez and the meteorological section airmen were moved from Seal Cove to Group headquarters to advise the flying control officers while the metmen remained at Seal Cove to brief

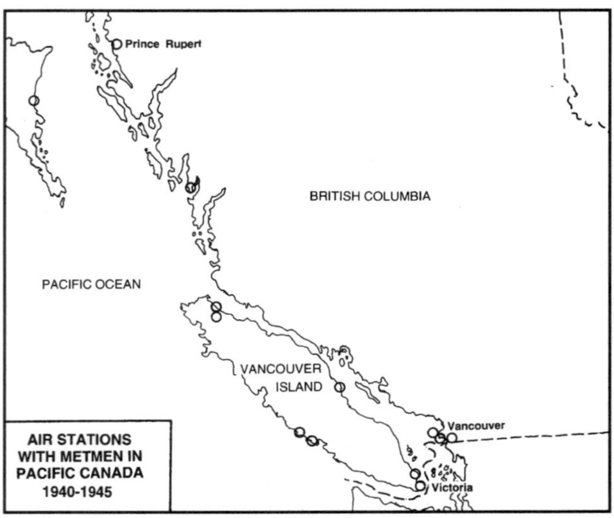

FIGURE 5

Air stations in Pacific Canada where metmen were stationed in wartime.

the aircrew. Ironically, in April 1944, at about the time a new landline teletype system began to operate, giving the section much needed additional observations and regional forecasts, the war situation had improved to the extent that the need for patrol work had greatly diminished. As a result, the Canso squadron was withdrawn and No. 4 Group headquarters closed. Gutierrez and most civilians departed in July 1944 but one metman and five airmen remained to serve itinerant Canadian and American military aircraft. In April 1946 the airmen observers departed and observations from Seal Cove ceased, but personnel at the Marine Radio station continued their synoptic observing program.

Metmen who served at Prince Rupert included J. Calder and J.P. Wilkinson. A Meteorological Assistant Grade 2, H.A. Mathison, came to Prince Rupert as a metman in late 1942 and remained until 1944.

NO. 2 GROUP HEADQUARTERS

In March 1943, when the WAC forecast office moved to Vancouver, the RCAF formed No. 2 Group headquarters at Victoria to control air force operations in the southern coastal region. In December, these responsibilities were increased as a result of an international defence agreement whereby the Group headquarters began to control all aircraft movements, both Canadian and American, civil and military, in southern British Columbia. The Group offices were in Belmont House, the building formerly occupied by the Command headquarters. Assisted by three metmen and the necessary map plotting staff of RCAF airmen, Don McIntyre (OIC) and Ken Harry were the meteorologists in the Group meteorological section for the year or so of its existence.

The chief duties of this office were to keep the operations room fully informed of weather conditions prevailing along the coast and to discuss with the RCAF controllers the current and forecasted weather situation as exhibited on the synoptic maps. Special upper wind and freezing level forecasts were also issued and tracings of the current synoptic map were provided to the controllers. Upper wind and other pertinent information were regularly furnished to the artillery and anti-aircraft batteries on Vancouver Island.

As the war situation improved National Defence decided early in 1944 to disband the Group headquarters. This occurred in March 1944 and the civilian meteorological staff was transferred to other offices. The metmen who served in this meteorological section included R. King, J.M. Quigley, and R.V. Maclean.

UCLUELET

The Ucluelet seaplane base, on the outer coast of Vancouver Island, was under construction before the war began and was occupied by the air force in September 1939. A reconnaissance squadron (No. 4) flying Blackburn Sharks and Stranraers arrived in June 1940 and flew patrols to identify and track shipping. The squadron was re-equipped with Cansos in late 1942. Patrols were discontinued in August 1944 and the RCAF left the base.

RCAF airmen began observations at Ucluelet in October 1940, the same month a metman arrived. The establishment was increased to two and then three metmen before patrolling ceased and the meteorological section closed in September 1944. Included amongst the metmen who served at Ucluelet were A.J. Filmer, H.I. Hunter, and E.J. Kermode. J.R. Taylor was a Meteorological Assistant Grade 2 who served as a metman at the base before the trained metmen arrived.

PATRICIA BAY

Construction of the Patricia Bay airport, just north of Victoria, was not completed when the war began but the RCAF began using it at once for operations and soon for operational training. At least eight different squadrons, both reconnaissance and fighter, were based at Patricia Bay over the war years. No. 111 was there as an army cooperation squadron in 1940, returned as a fighter squadron with Kittyhawks in 1942 and, after a period in Alaska, was again at Patricia Bay for the last part of 1943 before proceeding overseas with Curtiss Warhawks. In defence of Victoria and Vancouver, Squadrons Nos. 115, 132, 135, and 163 were there for various periods with Bolingbrokes, Kittyhawks, Hurricanes, and Venturas. For the last year of the war No. 133 was stationed there with Mosquitoes. A reconnaissance squadron, No. 8 was located at the base in the winter of 1944–1945 while No. 13, an operational training squadron, flew out of Patricia Bay in 1941 and 1942 with Venturas.

Patricia Bay was also home to Operational Training Units from October 1941 until the summer of 1945. No. 32 OTU, an RAF school, began flying at the base that October, and in June 1944 it became an RCAF school and moved to Comox. Another RCAF school, No. 3 OTU, trained aircrew at Patricia Bay from November 1942 until August 1945. To further congest the airport, Trans-Canada Airlines began to use the airport in 1943 for its scheduled service to Victoria.

A civilian observer began an observing program early in 1940 and RCAF airmen took over in July of that year. The first metman arrived that fall to serve the operational squadrons and later meteorologist Don McIntyre was sent to the station to organize a meteorological section to serve the RCAF operational squadrons and No. 3 OTU. Then another meteorological section was required for No. 32 Operational Training Unit and Fred Kelly organized it.

RCAF airmen plotted weather maps at both meteorological sections. Because of the long flights made in training and operations, meteorologists as well as metmen were required in both Patricia Bay meteorological sections. In March 1944 there were two meteorologists and four metmen in each section.

In 1944, after one OTU had departed and its meteorological section disbanded, the staff in the remaining meteorological section was increased to three meteorologists and five metmen. When the RCAF ceased flying training and all other flying from the base in September 1945 the second meteorological section was disbanded. Meteorological Division observers took over from the RCAF airmen and Patricia Bay became the Victoria civil airport late that year.

About twenty metmen and meteorologists served in the two Patricia Bay meteorological sections. Meteorologists Art Wright, W.R. Hamilton, and Bill Markham and metmen Dwight Williams, R.V. Poole, R.W. Smith, Ted Walker, W.S.C. Wallace, R.V. Maclean, M. Jenkins, F.M. Riddle, and J.T. Rush were in the section serving the operational squadrons for various periods. Meteorologists in the Operational Training Unit meteorological section were Don Cameron and D.M. Mitchell along with metmen J.E.A. Parnall, J.H. Wells, R.S. White, and J.G Bragg.

ALLIFORD BAY

Construction of the Alliford Bay RCAF seaplane base had progressed sufficiently to allow Squadron No. 6 to be stationed there in May 1940. The squadron flew Blackburn Sharks and Stranraers on patrol until April 1943 when Cansos replaced these aircraft. Squadron No. 7, with Cansos and Catalinas, replaced No. 6 in early 1944 and remained until July 1945.

RCAF airmen staffed the meteorological section from the time the station opened and in September 1941 were joined by a civilian observer. By 1943 communications had improved sufficiently to allow the reception of enough observations to warrant the plotting of weather maps and the posting in of a metman for map analysis and aircrew briefing. By June 1944 the staff was increased to three metmen at the station and that was the metman strength until the meteorological section closed in August 1945.

Metmen who were stationed at Alliford Bay during the war years were Don Kerr, N.E. Wolverton, R.W. Smith, J. Calder, and I.W. Findlay. E.G. Funnell, a Meteorological Assistant Grade 2, served as a metman until trained metmen were posted to the base.

BELLA BELLA

Construction of the Bella Bella seaplane base took longer than construction at the other coastal bases. It was not until December 1941, after a civilian observer had arrived in September, that a patrol squadron, No. 9, flying Stranraer aircraft, took up quarters at the base. The squadron was re-equipped with Cansos and Catalinas in May 1944 but the need for patrols diminished and the squadron was withdrawn later that year.

Regular meteorological observations, taken by a civilian Meteorological Division observer and three RCAF airmen, commenced in February 1942. With four observers on staff the office was open around the clock. While forecasts and observations from other stations were usually available by RCAF wireless from the WAC forecast centre, it was 1943 before the wireless teletype system brought a sufficient number of weather observations to enable the preparation of weather maps. At that time a metman arrived to analyze the maps and brief the aircrew. The 1944 landline teletype system, which greatly improved communications, arrived but a few months before activity at the base began to wind down. In August 1944 the station was closed as an operational base and the meteorological section disbanded.

Metmen who served at Bella Bella in wartime included I.W. Findlay, D.W. Ellis, J.H. Crowley, and J.T. Rush. Meteorological Assistant Grade 2 W.R. Brownlee was posted to the station early in 1942 to act as a metman.

COAL HARBOUR

The Coal Harbour seaplane base, near the northern end of Vancouver Island, was built for the RCAF during the early months of the war. The air force occupied the base in December 1941 and a reconnaissance squadron, No. 6, flying Cansos, was stationed there and remained until September 1945.

RCAF airmen took weather observations at this base from January 1942 until August 1945. The first metman arrived early in 1942 and by 1944 there were two on station to brief aircrew; they remained until the squadron left in 1945. Several metmen served at Coal Harbour including W.S.C. Wallace, Van Gordon, A.J. Ducklow, H.I. Hunter, D.W. Ellis, and E.J. Kermode. The first metman in 1942 was D. McKerricher, a Meteorological Assistant Grade 2.

TOFINO

This RCAF airport, built in 1940 near the Ucluelet seaplane base on the outer coast of Vancouver Island, was opened in the fall of 1942 and became home to a fighter squadron, successively one of No. 132 or 133 flying Kittyhawks and No. 132 with Hurricanes. Squadron No. 115, with Venturas, was there on patrol duty in 1944 and detachments from other squadrons were stationed there from time to time until June 1945. The RCAF kept the airport open until October 1946.

A metman arrived in the summer of 1942 to head the meteorological section, analyze and prepare weather maps, and brief aircrew. Later, in October, RCAF airmen began a regular observing program which they continued until the airport closed. There were two metmen on station in 1944 and both were posted elsewhere when the squadrons departed in the fall of 1944.

A list of the metmen who worked at Tofino includes R. McWilliams, Ted Walker, E.J. Kermode, Don Kerr, and G. Wilson.

BOUNDARY BAY

The Boundary Bay airport, built immediately southeast of the Vancouver urban area in 1940, was used for a year by a civilian operated Elementary Flying Training School when No. 18 was moved from Vancouver in December 1941. In October 1942, the RCAF took over the airport and used it as both a training base and as an operational base for fighter squadrons. That month, Squadron No. 133 arrived with Hurricanes, and No. 132 with Kittyhawks replaced this squadron during 1943–1944. Also, in 1943, Squadron No. 14 with Curtiss P-40 aircraft was stationed there prior to going overseas. The training unit was No. 5 Operational Training Unit flying Liberators.

Airmen at Boundary Bay began an observing program in January 1943 and continued until the RCAF closed the airport in September 1945. Meteorological services to station officials and aircrews were provided by metmen who arrived early in 1943 and remained until late 1944. For a period in 1944, because of the long flights by Liberators at the OTU, a meteorologist, W.H. Simons, headed the meteorological section. When Simons resigned later in 1944, Fred Kelly came to this section. Metmen who served on the station in wartime included T.A. Harrison, W. Kay, Jack Mathieson, Charley Hunt, D.M. Hanson, G.V. Adams, Bill Mackie, J.E. Parnall, and E.S. Weybourn.

PORT HARDY

Port Hardy airport, built for the RCAF near the Coal Harbour seaplane base at the north end of Vancouver Island, was opened in May

1943. A detachment of a composite squadron, No. 122, was stationed at the airport from May 1943 until April 1945. A reconnaissance squadron, No. 8, flying Venturas, arrived late in 1943 and remained until March 1944 when the need for patrols diminished. The RCAF left the airport in August 1945 when it was immediately taken over by the Department of Transport.

Because of the close proximity to Coal Harbour, where a meteorological section took a full set of meteorological observations, RCAF airmen did not begin observations at Port Hardy until February 1944. When the air force left the airport in the summer of 1945 civilian observers took over. One and then three metmen were at this station from 1944 until mid-1945.

The list of metmen serving at this station includes Van Gordon, A.J. Ducklow, H.I. Hunter, D.W. Ellis, and J.R. Wilson.

TERRACE

The Terrace airport, east of Prince Rupert in the interior, was built for the RCAF in 1943 as a facility for use in protecting the Interior Staging Route. This route was developed for transportation from Vancouver through Prince George to Prince Rupert. The airport opened in November 1943 and was occupied by Squadron No. 135 with Hurricanes until March 1944 and by Squadron No. 149, a torpedo bomber and then a reconnaissance squadron, over the same period. Transient military aircraft used the airport until it was closed in August 1945.

RCAF airmen took weather observations from February 1944 until the air force left in the summer of 1945. A meteorologist, Ken Harry, was at Terrace for the few months when the squadrons were there. There is no record of a metman ever being in the Terrace meteorological section.

JERICHO BEACH

The Jericho Beach seaplane base, located on English Bay in urban Vancouver, was a flying training base for the air force in the summers of the early 1920s. When the war began it was home to two operational squadrons, Nos. 4 and 6, flying Stranraer, Vancouver, and Shark aircraft, all obsolescent by that time. By June 1940 these squadrons had moved and it appears that no units flew on operations from this base after that month.

In the summer of 1939 it had been intended that the two men promised by the Meteorological Division to Western Air Command would be stationed there but in September they were posted elsewhere. During the war no metmen were ever posted to Jericho Beach but in 1943 the Western Air Command forecast office was located there when it moved from Victoria.

COMOX

The Comox RCAF airport opened in 1942 but it was the summer of 1944 before an Operational Training Unit flew from the base. At that time a Royal Air Force OTU at Patricia Bay was moved to Comox and became an RCAF training school. The meteorological section moved with the school to Comox and the staff remained largely the same. At Comox were meteorologists Don Cameron and D.M. Mitchell and metmen Parnall, Wells, White, and Bragg. Late in 1944 and 1945 W.S.C. Wallace and R.N. Chester were posted to the section.

SEA ISLAND

Sea Island airport opened in July 1931 and became Trans-Canada Airlines' Vancouver terminal when scheduled service began in 1937. That year, the Meteorological Division opened a District Aviation Forecast Office at the airport. In August 1940 the Department of National Defence took over the airport although it continued to operate as a civil airport. From the prewar period until November 1945 there was usually at least one squadron based at Sea Island. Besides reconnaissance and fighter squadrons, transport and communications squadrons were based there in the last two years of the war.

The existing District Aviation Forecast Office provided special forecasts for the RCAF patrol squadrons before and while the Western Air Command forecast office was being organized in the fall of 1940. The WAC meteorologists then took over forecasting for the squadrons.

The RCAF meteorological section at Sea Island was relatively small since the civilian Meteorological Division observers attached to the DAFO took the weather observations. There is no record of any meteorologists stationed at the Sea Island RCAF station but a metman, A.P. Edwards, was posted there in 1943 and 1944 to brief aircrews.

CHAPTER 11

ROUTE TO THE YUKON

As the result of extensive reconnaissance surveys in 1935 the Department of Transport recommended to government that a northwest airway be constructed between Edmonton, Alberta, and Whitehorse, Yukon Territory. By 1939 a few landing fields had been built along this route to accommodate an early air transport company, Yukon Southern, flying between Edmonton and the Yukon. That year, the federal government authorized the planning and construction of the airway with airfields at Grande Prairie, Fort St. John, Fort Nelson, Watson Lake and Whitehorse, and radio range stations at regular intervals along the route according to standard airway practice.

Construction began early in 1940 and later that year the Canada-United States Permanent Joint Board on Defence reviewed progress and strongly recommended that Canada complete the airway because of the great strategic value of such an inland route in the defence of Alaska and northwest Canada. The Canadian government undertook to pay all costs for the construction of the airports and the provision of ground facilities.

In September 1941, although construction was not yet complete, the airway, known as the Northwest Staging Route, became usable in daylight and fine weather. By the end of the year, radio range stations were in operation at the five airports which provided an airway to

Yukon and Alaska relatively free of coastal weather influences and the danger of enemy attack.[71]

PLANNING

In October 1940, the Department of Transport, responsible for all ground facilities on the staging route airway, informed the Meteorological Division of plans for this airway and asked for financial estimates for the necessary meteorological facilities. In accordance with the practice established earlier on the transcontinental airway, the radio range station operators at stations along the airway were to be responsible for the meteorological observations. In addition, on this airway, John Patterson proposed to post a metman (Meteorological Assistant Grade 3) to each of the five airports — Whitehorse, Watson Lake, Fort Nelson, Fort St. John, and Grande Prairie — to advise and brief itinerant pilots on the present and forecasted weather. The metman at these stations would also supervise radio operators in taking pilot balloon observations as well as a full program of surface weather observations.

Although a few telegraph and wireless synoptic weather reporting stations already existed at locations within a few hundred miles of the route, more off-airways stations were required to give some advance warning of bad weather to the metmen and to pilots using the route. Hudson's Bay Company posts at Fort Ware, Dease Lake, and Frances Lake were conveniently located and it was proposed that the managers at those posts could take the observations. The company had a record of good cooperation with the Meteorological Division and the posts already had radio equipment to transmit the observations. But, Frances Lake was the only station where this became possible and then only after a meteorological observer was posted there by the Meteorological Division to take most of the observations.

Also, an attempt was made to get the Royal Canadian Corps of Signals to have their operators at Dawson and Mayo, both in the Yukon, to increase the number of observations they took each day. Dawson and Whitehorse were connected to southern British Columbia on the Canadian Government Telegraph System but, since the telegraph offices were not open at night, it had become the custom for the RCCS to radio the Yukon observations south from Whitehorse to Prince Rupert for relay to the meteorological circuits. But the radio operators had no time to take more than their contracted two observations a day and so that attempt was unsuccessful.

During 1941, with the Canadian government giving prime attention to the needs of the war in Europe and the Department of Transport building airfields for the RCAF to use in the British Commonwealth Air Training Plan, work on completing the ground facilities at the northwest airway airports proceeded slowly. However, even without proper weather observations and forecasts, there was considerable Canadian and American commercial transport flying along the route. During this period the small meteorological office at Edmonton was staffed to provide but little weather information for pilots flying the route.[72]

ALASKA HIGHWAY

The Japanese attack on Pearl Harbor in December 1941 and the beginning of the war in the Pacific changed the national defence situation markedly in northwestern Canada. For air transport, the Canadian-built Northwest Staging Route to the Yukon was inland, away from danger of attack from the Pacific and, of considerable importance, where the weather was usually much better than that along the coast. Accordingly, this air route became essential for the defence of North America, especially for the United States Armed Forces who wished to use it for ferrying aircraft and for moving troops and supplies to Alaska. Also, the Americans had begun to provide new aircraft for the Union of Soviet Socialist Republics and these were ferried from Great Falls, Montana, along the route to Whitehorse where they were taken over by Russian pilots.

Until this time it had been very difficult to deliver supplies and building materials to complete the construction of the airports along the route. Beyond Fort St. John there were no roads and all supplies had to be transported to Fort Nelson by tractor train on snow roads in winter. Further north, supplies for Watson Lake were transported by ship to Wrangell, Alaska, by riverboat up the Stikine River and then overland on a temporary road to the airstrip. A road to roughly follow the northwest airway from the Peace River area of Alberta through the Yukon to Alaska was needed as a military artery, especially by the Americans.

In January 1942, the Permanent Joint Board of Defence advised that construction of the road should proceed and in February the United States government decided to build a highway capable of handling a flow of heavy traffic along the route. The Canadian government agreed in March and that month United States Army Engineers began the work. Called the Alcan Military Highway, a pioneer or "tote" road was completed from Dawson Creek, Alberta, to Alaska before the end of 1942.

During the next year, with contracts from the U.S. Public Roads Administration, civilian contractors, using Canadian and American labour, completed a permanent gravelled road. Access roads were also constructed to the airports along the route. The Canadian-built airway and the American-built road were complementary and opened up the northwestern part of the continent. After the war the highway was taken over by Canada and became known as the Alaska Highway.

REQUIREMENTS

In January 1942, American and Canadian meteorological officials met to consider what meteorological information would be required by the American military meteorologists in their support of flying operations in Western Canada. Of particular concern were the available meteorological facilities along the Northwest Staging Route. It was acknowledged that there were few observing stations in this sector of Canada and that the Meteorological Division had neither full-time employees nor pilot-briefing facilities north of Edmonton. Further, it was recognized that the existing small Edmonton meteorological office would be unable to provide the number of forecasts and the briefing facilities the Americans needed.

In response, the Meteorological Division took immediate action. Within a few weeks, a metman and two Meteorological Assistants Grade 2 (observers) were posted to each of the airports — Grande Prairie, Fort St. John, Fort Nelson, Watson Lake, and Whitehorse. The observers were to assist the radio range operators with the observing and the metmen to brief pilots. Further, it was decided to upgrade the Edmonton office to forecast centre status with an establishment of five meteorologists and sufficient assistant staff to perform the meteorological work. It was also decided that another full-service forecast centre would be established at Whitehorse when staff became available.

At this period in the war, while the Meteorological Division was still free to recruit qualified college graduates for training and work as metmen and forecasters, this was not the case with prospective weather observers and other technical and sub-technical staff. To recruit physically fit staff, permission had to be obtained from both the Civil Service Commission (CSC) and the local Mobilization Boards. Such fit men were required because of the severity of the climate and the conditions under which the personnel would live and work at airports on the route. There is no record of consideration being given to the employment of women in meteorological work on the staging route at this time.[73]

Despite the war and the urgent need for staff, CSC regulations had to be followed — advertising the positions, holding rating boards and interviewing the candidates, and establishing eligible lists. Each step, with the mandatory waiting periods, caused much delay before the positions could be filled. Following this process, the successful candidates in each group were gathered in Edmonton for a training course before being posted to the airports. This procedure became a recurring process during the war as more technicians were needed on account of resignations and staff expansion. In 1942, to help at the staging route airports before Canadian observers could be hired and trained, the USWB loaned eight civilian meteorological observers for posting to the Canadian stations to assist in observing. The RCAF flew the Americans in and out of the stations and their salaries were initially paid by the Weather Bureau. Later, the Meteorological Division repaid the USWB for the cost to that office for the American observers in Canada.

THE AMERICAN PRESENCE

Before Pearl Harbor, both Canadian and American air transport companies flew scheduled services from Yukon and/or Alaska along the new staging route to Edmonton. When flying north, the pilots visited the Edmonton office where they obtained forecasts and all available meteorological information regarding conditions at airports on the route. Early in 1942, the American military contracted with Northwest Airlines to transport equipment and military supplies along the route to Alaska and the company set up a civilian weather office at the Edmonton airport. Needing weather observations and forecasts, their officials had teletype drops connected to the Canadian meteorological circuits and they proposed to establish a forecast office with American civilian forecasters without realizing that both actions were in violation of Canadian regulations. The airline was offered Canadian meteorological services similar to those being successfully provided for United Airlines and other civilian contractors but Northwest claimed that they operated differently, making it essential that a forecaster be on hand at all times.

By this time the U.S. Weather Bureau was beginning to play a large role in the negotiations since John Patterson and his staff found it almost impossible to deal with the American military officers and with the Northwest Airlines officials. Meetings were held at Edmonton at which John Patterson, Tom How, the officer-in-charge of the Edmonton office, met with USWB personnel, airline officials, and American military meteorologists and proposed some solutions. There, it was agreed that all available Canadian and American forecast personnel be pooled at the Edmonton Meteorological Division office and that teletype facilities be installed to connect the expanded forecast office with Northwest Airlines, the US Army Air Force, and the radio range stations at airports along the route.

Shortly, however, the Chief of the United States Army Air Force ruled that the USAAF must have its own meteorological staff at all airports and so the combined office did not materialize. Further, this meant that "pooled staff" offices could not be set up at any airport along the route. However, facilities were put in place to allow the Edmonton office to provide meteorological observations and Canadian forecasts from the Canadian circuits to the American civilian airline dispatchers. It should be noted that while the Canadian government would not allow the establishment of American civilian forecast offices in the country, the North American continental defence agreement gave full authority to the American military forces to establish observing and/or forecast offices manned by military personnel in Canada.

MILITARIZATION

By June 1942, the extensive needs of the American military for meteorological services along the route were being realized. The US Army required teletype drops on the Canadian circuit at all their stations and permission was requested to allow the coding and transmitting of all Canadian reports and forecasts on their own radio facilities. The USAAF established forecast offices at Whitehorse, Watson Lake, Fort Nelson, Fort St. John, Grande Prairie, and Edmonton to provide forecasts and briefings for their ferrying, transport, and communications flights. Plans for an office at Regina never materialized.

It should be noted that the American military forecast offices were staffed differently than Canadian forecast offices. While a "forecast office" in wartime Canada implied a staff of independent forecasters with graduate meteorology training, senior non-commissioned officers, usually with fewer qualifications and less training, staffed most USAAF forecast offices in Canada.

On July 1, 1942, the United States Army Air Force militarized all American activities along the staging route and there were no American civilian meteorological personnel working in Canada after that date. Canada retained responsibility for the flying aids along the staging route, which included operating the radio range stations and taking weather observations. With greatly increased air traffic it was decided, at about the same time, to add five intermediate airway range stations to the system. These were built by the Department of Transport at Snag, Aishihik, and Teslin in the Yukon, and at Smith River and Beatton River in British Columbia. In 1943 and early 1944 the Meteorological Division posted four observers to each radio range station where hourly and synoptic observations were taken and transmitted.

It soon became clear that the Canadian Meteorological Division would not be able to provide the additional airways and off-airways observing stations required by the Americans. Accordingly Canadian government authority was given to the USAAF to establish additional stations along the Northwest Staging Route and at other places throughout northern Canada as thought necessary. (By 1944 there were twenty-five USAAF observing stations along the staging route and as many as thirty-five more at other locations.)

Besides those along the staging route, other American forecast offices were set up at Prince George, to cover the interior British Columbia route, at Norman Wells in connection with the Canol Project, at Calgary and Lethbridge to service transport aircraft and ferrying operations into Edmonton from the United States, and at The Pas, Churchill, and Coral Harbour along the Crimson Route. Further mention is made in Chapter 7 of other USAAF forecast offices established on Baffin Island and in Quebec and Labrador in connection with the northeast ferrying route and in Chapter 18 of the American observing stations in the North.

At the American forecast offices along the Northwest Staging Route the usual USAAF staff consisted of three forecasters and six observer-plotters who decoded the incoming weather data, typed the data in airways sequence form, and delivered the information to the United States military users. At the Canadian meteorological sections at the same five airports the staff consisted of one or two metmen for aircrew briefing duties and three civilian observers who observed and coded the weather reports for radio transmission and supplied information to air and ground units requiring it.

During 1943 it soon became obvious to both the Canadians and the Americans that the Edmonton municipal airport, used by military and civilian aviation from both countries, would soon be excessively overcrowded. Accordingly, authority was given to the Americans to build a new military airport at Namao, a dozen or so kilometres north of Edmonton. In 1945 the airport was completed and the Americans moved their transport flying operations and the ferrying of new aircraft to the USSR via Alaska to the new airport. After the war the Namao airport was taken over by the RCAF.

OBSERVATIONS

The Canadian meteorological service was responsible for the hourly and synoptic weather observations at the five main airports and at the five intermediate airfields although there were radio range operators at or near each airfield. The Canadian observers at the five main stations were also responsible for coding the reports for radio transmission and for supplying copies to the various units concerned. At some stations, American observers assisted in the Canadian office by decoding incoming reports and typing them in airways sequence form for distribution. Unfortunately, the Canadian and American agencies used several different communications systems for the distribution of these reports and this led to problems which are described below. A more complete account of observations along the staging route is given in Chapter 18.

FORECASTS

By November 1942, well before the Whitehorse forecast office was opened, the Edmonton office provided route and terminal forecasts every six hours for the entire Edmonton to Whitehorse route. Each was valid for eight hours with a future outlook for eight hours. The forecasts were sent out over the Radio Division's wireless system to the main airports where metmen were stationed to brief pilots on request. Regular forecasts were also issued for the air route from Prince George in interior British Columbia to the staging route at Fort St. John and for flights over the Edmonton-Penhold-Regina-Minneapolis route for American civilian air contractors. During construction of the Alcan Highway, forecasts were issued each day for the American civilian contractors for whom temperature forecasts were of great value.

After the Whitehorse forecast office became fully operational on a twenty-four-hour basis, in March 1944, that office became responsible for terminal and route forecasts from the Alaskan border through Whitehorse to Fort Nelson and for an air route from Fort Nelson to Norman Wells. The Whitehorse office briefed pilots of the United States Navy, the RCAF, two civilian airlines, and airlines under contract to the American navy and army.

While many terminals and air routes can claim individual and special adverse flying weather problems, the Northwest Staging Route had a rather unique one. This was the problem of "upslope weather" which was of particular importance in the British Columbia portion of the route. The continual movement of weather systems from the Gulf of Alaska eastward over the route could cause enough adverse weather but when combined with the southward and westward flow of cold air behind a cold front, very severe weather with low ceilings and restricted visibility occurred. The problem of forecasting such conditions for propeller-driven and relatively low flying aircraft was significant.[74]

COMMUNICATIONS

When construction was begun on the Northwest Staging Route meteorological teletype facilities did not exist north of Edmonton. The synoptic reporting stations in the Peace River country used the telegraph system to send observations to Edmonton but the observations taken by Royal Canadian Corps of Signals radio operators at stations further north in the Mackenzie River basin were sent south by RCCS short wave radio transmission. Observers at stations in northern British Columbia and the Yukon were able to use the existing Canadian Government Telegraph System to send out their weather observations although, increasingly, the RCCS wireless system between Whitehorse and Prince Rupert was being used since it was more reliable and operated at night.

Along the route, at first, the Radio Division's short wave radio system was the only method of communication and it was used to collect observations from the airports, to distribute the observations, and to transmit the Edmonton forecasts. When the Americans came to the staging route a local teletype circuit was set up in Edmonton connecting the forecast office with the United States Army Air Force headquarters, Northwest Airlines for their contract flying, and the RCCS for the receipt of reports from the north. Shortly the USAAF installed additional teletype circuits connecting their offices in Edmonton with Spokane, Washington, and a radio circuit with Annette Island, Alaska, for the exchange of data and for the receipt of observations from Alaska and the Yukon via Whitehorse. As the Americans occupied the airports along the route the US Army and USAAF installed new and different radio circuits.

Establishing and maintaining efficient and effective communications systems remained a problem until the United States government erected a telegraph line from Edmonton to Fairbanks, Alaska, along the Alcan Military Highway. Construction of the landline began in the fall of 1942 and reached Whitehorse in May 1943. The United States Army Signal Corps became responsible for maintaining the line where one duplex circuit was dedicated to meteorology and was to be used jointly by the Canadians and the Americans. Teletype machines were located in both the Canadian and American forecast offices at Edmonton and Whitehorse. But at each of the other primary airports, where the Canadian offices took and transmitted the observations, the teletype machines were to be in their offices. However, within a short time after the circuit became operational the American offices demanded and obtained teletype machines as well.

Since no landline connections were made with the intermediate radio range stations at Snag, Aishihik, Smith River, Teslin, and Beatton River, these stations had to communicate on the Department of Transport radio system until the end of the war. Teletype machines were placed in the American teletype repeater stations along the route where military personnel took the observations.

The teletype circuit was not too efficient at first and the original radio schedules were maintained to duplicate the traffic until the teletype circuit became reliable. Because of the length of the circuit, fully half of each hour had to be devoted to the collection and relay of the hourly reports. Synoptic, pibal, and radiosonde observations were scheduled in the other half along with forecasts and the relay of observations from the United States which was required by the Americans for their long distance flights. To help in co-ordinating the communications system and the provision of services in general, a United States Weather Bureau liaison officer was stationed at Edmonton in 1942 until April 1943 which was of great assistance to the officer-in-charge of the Canadian office at Edmonton.

PROBLEMS

With several jurisdictions involved and generally poor communications, there were numerous complaints from pilots about the service the Meteorological Branch, and the Department of Transport in general, provided to them. Canadian Pacific Airlines, which maintained a scheduled air service along the route, complained that the observers did not monitor the weather carefully enough, that the code used did not give their pilots all the meteorological information desired and, furthermore, their pilots did not understand the code used in the broadcasts. The Meteorological Division pointed out that CPAL pilots never came to a weather office for briefing before flights, did not understand the meteorological terms in use, and refused to cooperate by not completing pilot report forms on the in-flight weather. The USAAF pilots filed pilot report forms fairly regularly but omitted much information.

In February 1943, the United States Army, responsible for building and maintaining the highway, wanted meteorological observations from thirteen of their service stations which were generally located every fifty miles (eighty kilometres) along the road. It was agreed that these stations should be entirely equipped and manned by the United States. At this time there were as yet no observations from the teletype repeater stations but the Edmonton forecast centre provided the American contractors with weather and temperature forecasts which they used for scheduling construction. The contractors needed not only forecasts but also records of the actual weather for possible use in settling disputes with the US Army regarding completion dates.

Much confusion and unnecessary duplication of work took place during this early period. Three United States military services, the army, the air force and the navy, were involved and frequently made changes to their communications operating systems without notifying the Canadians. At Whitehorse, the USAAF caused some confusion when they began a radio interception and transmission of the Alaskan weather reports, a task that had been originally assigned to and was being done by the Radio Division of Transport. Relations between the Canadians and the USAAF forecasters at several airports became very strained as the latter blamed the Canadians for not supplying complete observations on time or of not taking the necessary special observations. In April 1943 the USAAF Washington headquarters ordered their forecast offices to begin taking duplicate observations as a check on the Canadians.

Most of the difficulties arose from the fact that the weather observations taken by the Meteorological Division observers were given to various users by five different agencies, sometimes after coding and decoding which often ensured errors and omissions. The agencies involved were:

a) Radio Branch of Transport — Using Alaskan scramble code, each radio range station broadcast the hourly and special sequences for that station, and for the adjacent stations, to all aircraft flying the range. Each hour the range operator was supplied with a typed copy of the relevant observations, which had been checked by the metman, to ensure that the latest and correct information would be given to pilots who asked for it.
b) Royal Canadian Air Force — At each station the RCAF radio station broadcast landing instructions to aircraft approaching the airport. Included in the instructions was local weather information taken from typed copies of the sequences provided by the observers.
c) United States Army Air Force Ferry Command — A radio station at each airport was supplied with a typed copy of all weather reports received and these were broadcast in code to Ferry Command transport aircraft on request.
d) United States Army Air Force — Typed copies of the observations were given each hour to another air force radio station that broadcast hourly reports in Alaskan scramble code to any USAAF aircraft — fighters, bombers, or transports — requesting the information.
e) USAAF forecast office — A typed copy of all hourly weather reports was given to the USAAF forecast office for their use in forecasting.

Attempts were made in May 1943 to list the responsibilities of the various weather offices and radio stations with a view to eliminating duplications and cutting down on errors. While the work within the Department of Transport's Meteorological and Radio Divisions was centrally controlled and well coordinated, this was not the situation with the military radio stations serving the pilots. No definite action was taken as the agencies involved thought that time would iron out many of the difficulties especially when teletype communications replaced much of the radio traffic.

ADMINISTRATION

On July 1, 1942 all American personnel and services on the staging route were militarized and in mid-October the Royal Canadian Air Force took over complete control of the airports including maintenance and operations. However, as was done at the RCAF stations in southern Canada where aircrew were trained under the British Commonwealth Training Plan, the Department of Transport's Civil Aviation, Radio and Meteorological Divisions served the air force with civilian metmen and radio range operators. Some changes were made later, however, when the RCAF took over operation of all control towers on the route from Transport.

ACCOMMODATION

When the first meteorological personnel were posted to the radio range stations on the Northwest Staging Route early in 1942 the Department of Transport was responsible for providing living accommodation. Earlier, on the transcontinental airway from Vancouver to Moncton, only the Radio Division personnel needed accommodation at isolated stations since no meteorological men were posted there. When dwellings were built by the department these were assigned to individual radio operators by the Radio Division authorities.

At the five isolated airports on the staging route, where Radio and Meteorology employees shared weather-observing duties, accommodation became a serious problem for those in meteorological work. Although it had been decided at Air Services headquarters that the Meteorological and Radio Division staffs would share the accommodation, the latter arrived first and were given the available accommodation. Locally, the radio people were in charge and they were loath to make any changes which would put radio people at a disadvantage which meant that the arriving meteorological staff had to take what was left or search for something off station. Because of this, relationships between the Radio and Meteorological Division staffs were never very satisfactory.

When the decision to establish five intermediate range stations was made in July 1942 it was decided that the Meteorological Division observers should be solely responsible for weather observations and arrangements were made for their housing. After October 1942, when the RCAF took over complete control of the airports, some more buildings were constructed at the main airports. At stations where the RCAF maintained an officers' mess and quarters it was arranged that meteorological personnel could be accommodated there. But, throughout the duration of the war, the level of accommodation and messing arrangements along the route remained uneven.

It is interesting to note that when the meteorological observers and metmen first went to these isolated stations, living conditions were very primitive and little attention was paid to personal appearance. With the arrival of the USAAF and RCAF servicemen in uniform, and increasing activity at the airports, the appearance of the meteorological staff began to have an adverse affect on the Meteorological Division and necessitated a sharp reminder to all from John Patterson of the necessity to improve. Also, as other organizations began to arrive on the route, it became obvious that the meteorological personnel were not enjoying some of advantages given to others for serving in isolation. One of these was the payment of transportation charges on annual leave to either Edmonton or Vancouver; such charges were later authorized by an order-in-council.

EDMONTON

In 1937 a meteorological office was established at the Edmonton airport (Blatchford Field) in anticipation of scheduled flights by Trans-Canada Airlines. The field had been opened in 1927 and was primarily used by the local Aero Club for flying training and for private flying. In 1930 and 1931 it was used as a terminal for the Western Air Mail Service connecting Winnipeg to Edmonton and Calgary. The field had also been used extensively as a base for bush flying in which surveyors, prospectors, and miners were transported with their supplies into isolated Mackenzie and Peace River areas.

Edmonton had been an important telegraph weather reporting

station since 1880. In 1937 two Meteorological Division observers commenced observing at the airport and meteorologist Tom How arrived in August 1938 to provide services to TCA when flights to Edmonton were commenced. At about this time Edmonton became one of the few apobs (aeroplane observations) stations in Canada where temperature, pressure, and humidity profiles were measured daily to heights of about ten to twelve thousand feet using a small aircraft.

In August 1940, the RCAF located a civilian-operated Air Observer School, part of the British Commonwealth Air Training Plan, at the Edmonton airport. Meteorologist Fred Kelly arrived at that time to establish a meteorological section and provide a lecture course for the student aircrew. Using Lethbridge DAFO forecasts, Kelly also briefed the civilian pilots before training flights. Meteorologist Syd Buckler followed him and later, when the Edmonton forecast office had opened, metmen took over the briefings and student lecturing.[75]

To meet the needs of flying along the Northwest Staging Route and down the Mackenzie Air Route the Edmonton forecast office was opened by the summer of 1942 with Tom How as officer-in-charge. Meteorologists Allin Jackson, Don Currie, Stu Dewar, and A. Gibb arrived in the summer months and remained at that office for the duration. Harry Tucker and Carl Mushkat were also at this office for most of 1942 before moving to other offices. Other meteorologists at Edmonton for parts of 1944 and 1945 were E.A. Johnson, George Robertson, Don Slater, and A.E. Allison. In the fall of 1943 the office began issuing public weather forecasts for Alberta.

WHITEHORSE

In September 1940, a Royal Canadian Corps of Signals station at Whitehorse began taking and transmitting two synoptic observations each day and continued until a weather reporting station, manned by Meteorological Division observers, was opened at the Whitehorse airport in April 1942. With observations from airports along the Northwest Staging Route and forecasts from the Edmonton forecast office, Harlan Thompson, then an experienced observer, and assistants at the Whitehorse office provided briefings and information for Canadian and American civilian and military aircraft involved in local flying and flights along the airway. Until the office could be properly staffed the United States Weather Bureau loaned an observer and United Airlines provided a forecaster. Then, in April 1943, a fully staffed USAAF forecast office opened and subsequently served all American aircraft passing through.

Meteorologist Hugh Cameron was posted to the Canadian office in the summer of 1943 and with the arrival of other meteorologists — George Legg, Clarence Thompson, Ken Harry, and Burn Lowe — in the spring of 1944 the office was able to provide a twenty-four hour service. Gradually the establishment increased to five forecasters with the return of Harlan Thompson, now a meteorologist, and the addition of Charlie Goodbrand and E.A. Johnston. During this period Clarence Thompson and Cameron departed. Besides meteorologists the office establishment consisted of two teletypists and eight observers. When Harry and Lowe departed in 1945 Ted Walker and Don McIntyre came to replace them. There were six meteorologists on staff at the end of that year.

Amongst the pilots briefed were those of the United States Navy; Pan American and Canadian Pacific Airlines, both under contract to the American military services; the RCAF; scheduled Canadian Pacific flights; and Alaska-Star Airlines. The Whitehorse forecast office was responsible for issuing forecasts for both the northern part of the staging route and for the Canol Project route.

With construction and especially after traffic began along the Northwest Staging Route, accommodation was very hard to find in Whitehorse. When a house became available for purchase early in 1942, it was purchased by Transport for the meteorological personnel. Located a mile or so from the airport the small staff occupied it in April and made their own arrangements for meals and caretaking. When the Department of Transport constructed a thirty-two-man barracks the Radio Division claimed most of the space although five rooms were eventually allocated to the Meteorological Division.

OTHER METEOROLOGICAL SECTIONS

Archives do not reveal the names of all the metmen that were posted to the four main airports between Whitehorse and Edmonton in wartime. However, it is known that early in 1942 E.A. Johnston arrived at Fort Nelson, Lloyd Richards at Fort St. John and Don Slater at Grande Prairie. Within a year or so each of these metmen was brought back to Toronto for the advanced course to qualify as forecasters and were replaced by other metmen. Stu Shannon was one of those; he was at Fort St. John for a year after the summer of 1944.

WINDING DOWN

In 1945, with the end of the war in the Pacific, the number of USAAF flights along the staging route dropped rapidly. Since the Americans proposed to abandon most stations by October 1, 1945, meetings were held that summer regarding the continuation of services. The USAAF agreed to keep their offices in Whitehorse, Fort Nelson, and Edmonton functioning until the Meteorological Division decided what stations and services should be maintained. From December 1945 until March 10, 1946, when the USAAF vacated their base at Whitehorse, the American forecasters used the Canadian analyses and forecasts to provide service for their military aircraft.

Both the Edmonton and Whitehorse forecast offices remained open, staffed with meteorologists, after the war but the metmen at the other airports on the staging route were withdrawn.

[71] Details and dates regarding the development of the airway and airports are given in T.M. McGrath, *History of Canadian Airports* ([Ottawa]: Lugus, 1991). See also *Canada Yearbook*, 1945, pp. 706–709.

[72] Personal archives, letter from George Robertson, a meteorologist who began his career as an observer at Edmonton in 1938, 2 May 1981.

[73] About 100 Canadian and American women were employed for office work at the Fort St. John headquarters of the United States Public Roads Administration and a few in the construction camps. See Phyllis Lee Brebner, *The Alaska Highway — A Personal and Historical Account of the Building of the Alaska Highway* (Erin, Ontario: Boston Mills Press, 1985).

[74] L.L. Kolb and M.M. Goodmanson, "Upslope Weather Along the Inland Alaska Air Route," *Journal of Meteorology*, vol. x, nos. 3 and 4, December 1944, pp. 98–108.

[75] Robertson letter, May 1981.

CHAPTER 12

THE OIL ROUTE

After Pearl Harbor, the United States government threw almost unlimited resources into the defence of Alaska. Besides building the Alcan Military Highway and improving the existing meteorological and telecommunication services along the Northwest Staging Route, the United States undertook the Canol Project. This included drilling new oil wells at Norman Wells, building a pipeline over three mountain ranges, building a refinery at Whitehorse, and constructing airfields along the Mackenzie air route. (Canol is an acronym for Canadian Oil.)[76]

The development of meteorological services for the Canol Project and its associated Mackenzie air route was unique in the history of meteorological services in wartime Canada. Meteorological observations were needed from areas far to the north of existing landlines with normal teletype communications and forecasts were needed for areas for which both the Canadian and the American meteorologists had had little or no forecasting experience. For these reasons the meteorological services provided at first were somewhat like those provided in southern Canada for airmail flights in the 1929 to 1932 period when weather observations were probably more important to the pilots than forecasts. It is ironic that, by the time sufficient resources, personnel, and experience had been brought to the Canol-

Mackenzie air route meteorological system to enable the provision of observations and forecasts somewhat similar to those available along the transcontinental airway, the Pacific war was over and there was no more need for such an extensive and expensive meteorological system in that part of Canada.

CONCEPT

The Canol Project was the name given to a joint Canadian-American defence undertaking developed after Pearl Harbor to guarantee a supply of oil to the large military force which was expected to move into northern Canada and Alaska and the adjacent seas. Canada provided the sites and rights of way while the Americans organized and paid for the construction and most other costs of the project.

To bring the Mackenzie valley oil wells into production and to build and operate a pipeline from Norman Wells to a new Whitehorse refinery meant that extensive air, water, and land transport routes had to be built and operated. This included building a road parallel to the pipeline from Whitehorse through Macmillan Pass and over mountainous country to Norman Wells. Much of the equipment and supplies for the endeavour had to be brought north from Edmonton on rivers of the Mackenzie River system during the open water season and by tractor train in winter.

Besides, extensive use had to be made of air transport in winter to take construction supplies to the project. To do this the contractors required weather observations and forecasts, as did the USAAF in case air defence was required for the project. To meet these needs the Mackenzie air route was organized and a string of airports built at a number of settlements between Fort McMurray and Norman Wells. While the United States government built all the facilities and American aircraft were expected to do most of the flying along the route, Canada was responsible for the basic meteorological services.

OBSERVATIONS AND FORECASTS

In the mid-1920s the Royal Canadian Corps of Signals had established wireless stations in the Mackenzie basin and part of their duties was to observe and report the weather for the Meteorological Service. Aklavik, Fort Simpson, and Fort Smith began reporting twice a day in the mid-1920s and Fort Norman and Fort Resolution a few years later. There were no scheduled forecasts, either public or aviation, for the Mackenzie basin until the time of the Canol Project, except for attempts made at the Edmonton meteorological office to give general weather advice to aircraft flying down the Mackenzie.

In October 1942, to provide meteorological services for air transport, the Americans asked for hourly reports from twenty-four specific locations in northwestern Canada, mostly at locations without any kind of an observing station. The Americans proposed, and Canada agreed, that the United States Weather Bureau would supply the necessary instruments and equipment for the needed stations, that the Alaskan Communications System, then being developed, would provide the necessary communications facilities, and that the USAAF would post three trained observers to each site. In turn, the Edmonton forecast office of the Meteorological Division would collect and decode all synoptic and hourly weather reports before making them available to the USAAF.

At four of the stations named — Fort McMurray, Fort Smith, Fort Simpson, and Fort Norman — the RCCS wireless operators already furnished four observations a day. However, these RCCS sites were located at considerable distances from the airports then under construction and aviation requirements called for twenty-four observations a day at the airports. Accordingly, it was agreed that the wireless stations would continue taking four synoptic observations a day and that the USAAF would establish observing stations at the airports where hourly observations would be taken twenty-four hours a day. The USAAF began observations at these stations by June 1943 when they established a forecast centre at Norman Wells and set up radiosonde observing stations at Fort Smith and Norman Wells.

With the Canol project well underway, meetings were held in Ottawa in the summer of 1943 to consider establishing a new airway from the Mackenzie route across the top of the Yukon and Alaska for the delivery of aircraft to the USSR. However, the planning never became sufficiently advanced to consider what meteorological services might be required.

Although there was a great deal of flying for the Canol Project in the summer of 1943 and despite the new observing stations, the meteorological service remained quite inadequate largely because of poor communications. The American agencies did not have a unified system and there was an apparent lack of co-operation between them and the RCCS wireless people. The Alaskan Communications System, employed by the Americans along the Mackenzie route, used short wave radio but the system was very erratic. It was never reliable enough to deliver a sufficient number of observations to the Edmonton office to allow the forecasters to issue useful forecasts for pilots and dispatchers.

However, the RCCS stations had both long and short wave radio and by using their long-range transmitters were able to work directly with Edmonton, although relays at Fort Smith were sometimes necessary. By this means, in time, enough observations reached Edmonton to allow weather map analysis and the issuing of the useful forecasts for the Mackenzie airway.

In December 1943 the RCCS at Fort Nelson acquired additional equipment and personnel to enable that office to provide a twenty-four-hour wireless service with Norman Wells, Fort Norman, and Fort Simpson. Long-range transmitters at Fort Norman enabled the operators to work directly with Fort Nelson and improve the exchange of observations between the Northwest Staging and the Mackenzie air routes compared to what had earlier been the case. Early in 1944 the new Whitehorse forecast office became responsible for forecasting for the Whitehorse to Norman Wells route along the Canol road and a new RCCS wireless station was established at Fort Good Hope in July 1944.

PROJECT ABANDONED

The pipeline from Norman Wells to Whitehorse was completed and ready for operation early in 1944 and the refining of oil at Whitehorse commenced that spring. Then, in June, when the Japanese threat to Alaska and Pacific Canada had decreased markedly and the construction phase of the project completed, the Meteorological Division was advised that the United States Army Air Force planned to withdraw their meteorological personnel from the Canol stations that fall. However, oil was pumped through the pipeline and the Whitehorse refinery continued to operate until April 1945.

The USAAF planned to continue to operate aircraft for this period over the Canol and Mackenzie routes after their personnel were removed from the weather offices. To meet the pilots' needs, the Meteorological Division was asked to take over the weather reporting stations and to staff the forecast office at Norman Wells. The Canadian government agreed that communications would become the responsibility of Canada with the RCCS taking over the work except at radio range stations where the Radio Division would be responsible. The Americans agreed to leave all instruments and equipment at the stations they vacated subject to later financial settlement between the governments. Included in this transfer was a teletype circuit from Whitehorse to Norman Wells along the Canol Road but the circuit was far from dependable; it was ultimately discontinued and wireless communications used for weather information.

Not all observing stations established by the Americans were needed by the Meteorological Division. It was considered sufficient to operate seven Mackenzie air route stations to take hourly and synoptic observations, five pilot balloon stations, and two radiosonde stations while the Norman Wells forecast office would operate as long as the United States Army Air Force needed services. Since the Canadian service could not immediately staff the observing stations, the United States Weather Bureau sent in civilian observers for several months. The Canadian delay was caused by the inability of the Meteorological Division to recruit a sufficient number of physically-fit men willing to take these northern postings since, at that time, the Mobilization Boards were very reluctant to grant deferments to any men subject to call-up for military service.

At the former USAAF airport observing stations at Fort McMurray, Fort Smith, and Fort Simpson, the Meteorological Division observers had taken over by January 1, 1945, while the RCCS provided the staff for communications and weather observing at Embarras, Fort Resolution, Hay River, Fort Providence, and Wrigley.

ROYAL CANADIAN AIR FORCE

It is noteworthy that, in contrast to the situation in other parts of the country, the Royal Canadian Air Force required few services along the Canol and Mackenzie air routes. As along the Northwest Staging Route, the Mackenzie route airports were under the control of the Department of Transport until October 1942 when the RCAF took over complete control.

The RCAF created a separate Northwest Air Command, based in Edmonton, in August 1944 but it was disbanded at the end of the war. No operational flying units were assigned to the Northwest Air Command. Transport Squadrons Nos. 164 and 165, flying Lodestars and Dakotas, stationed detachments at Edmonton for periods from 1943 to 1945 and aircraft from communications squadrons flew staff officers and other officials over the Canol and Mackenzie routes. Although most RCAF Commands had forecast offices staffed with Meteorological Division civilians, no request for such an office was received from the Northwest Air Command. Instead, observations, weather maps, and forecasts were provided to the Command's operations office by the Edmonton forecast office.

NORMAN WELLS

In December 1944 the Meteorological Division posted three metmen and four assistants to the Norman Wells office along with two radiosonde technicians to gain experience before the USAAF staff withdrew. Jeff Quinn was one of the metmen posted to Norman Wells in late 1944; there is no surviving record of the names of other metmen who must have departed from the Meteorological Division in 1945. Nor is there a record of any meteorologists being posted to Norman Wells indicating that the metmen depended on the Edmonton and Whitehorse forecast offices for guidance in weather map analysis and forecasting.

The needs of the USAAF for meteorological services rapidly decreased as the war in the Pacific was drawing to a close. The Norman Wells meteorological office closed in August 1945 although the surface weather observers and the radiosonde technicians remained and continued their observations. (During the winter of 1945–1946 the Norman Wells office participated in Exercise Muskox by providing forecasts for the RCAF as it flew supplies to the ground party moving through the Mackenzie region.)

The USAAF did not have forecast offices at any other airports along either the Canol route or the Mackenzie air route. The RCAF never requested meteorologists or metmen at other Mackenzie air route airports except for Fort Smith where Alf Ingall was in 1945 for two months.

[76] Richard S. Finnie, "The origin of Canol's Mackenzie air fields," *Arctic*, vol. 33, no. 2, June 1980, pp. 273–279.

CHAPTER 13

AIR TRAINING PLAN

The British Commonwealth Air Training Plan is considered by some historians to have been Canada's major contribution to the Allies' victory in World War II. Politically, in the early war years, the training program was in some aspects the most acceptable form of assistance Canada could provide since it did not involve sending a large army to serve overseas, an army most likely raised by conscription. At home, the plan demanded a major effort in resources and manpower and this stimulated the Canadian economy and helped bring the country out of the great economic Depression of the 1930s. And without doubt, the aircrew trained in Canada did have a marked influence on the war in Europe.[77]

The need for flying weather forecasts and instruction in meteorology for the British Commonwealth Air Training Plan (BCATP) students quickened the development of aviation meteorology in Canada. Earlier, in the mid-1930s, the national meteorological service had been given resources to modernize and expand as part of the government's response to the meteorological needs of commercial aviation. Then, during the first years of the war, and largely because of the needs of the RCAF in implementing the BCATP, this meteorological expansion accelerated rapidly. By 1945, nearly 400 new weather forecasters and metmen had been trained to provide service for the Royal

Canadian Air Force and Canada's national meteorological service was about ten times larger than it had been in 1939.

Much has been written in recent decades about the British Commonwealth Air Training Plan. Its history is the subject of Hatch's *Aerodrome of Democracy* and a major portion of Douglas's *The Creation of a National Air Force*. Further, the BCATP is the subject of Williams' *The Plan*, Conrad's *Training for Victory*, and Dunmore's *Wings for Victory*.[78]

Unfortunately for those interested in the history of Canadian meteorology the words "meteorology" or "weather" do not appear in the indexes of any of these books. Of course weather does enter into some stories dealing with accidents, delays in flying training, and as a subject in ground school for pilots and observer/navigators. But nowhere is there any mention of the civilian metmen and meteorologists who provided the weather forecasts and meteorological instruction.

THE AGREEMENT

On December 17, 1939, four British Commonwealth nations — the United Kingdom, Canada, Australia, and New Zealand — reached agreement on a plan for training aircrew to serve in World War II. Under the terms of the British Commonwealth Air Training Agreement Canada became responsible for conducting the training and for command of the people and bases involved. Described as a co-operative "undertaking of great magnitude" the document was officially called an "Agreement Relating to the Training of Pilots and Aircraft Crews in Canada and Their Subsequent Service."

As already described in Chapter 3, representatives of the commonwealth nations had begun meeting in Ottawa in mid-October 1939 to consider setting up a flying training plan. To reach agreement it was necessary to solve several difficult political and financial problems that would arise from the plan. It was eight weeks before this could be accomplished on the night of December 17, 1939. Prime Minister King then addressed the Canadian people in a national radio broadcast and told them, "[The plan] will establish Canada as one of the greatest air training centres of the world. Its development will result in a rapid increase in the number of air training schools in the country, and will achieve a steadily increasing output of highly trained pilots, observers and air gunners. . . ."[79]

FLYING TRAINING

The Canadian Air Force had come into being in March 1923 and became the Royal Canadian Air Force in April 1924. An Air Training Command was established in October 1938 and, after the BCATP agreement was signed, the Command was divided into four numbered commands with headquarters in Toronto, Montreal, Winnipeg, and Regina (moved to Calgary in 1941). Each command was responsible for all BCATP training in its geographical area using national standards.

Each Service Flying Training School was commanded by an RCAF officer with the rank of Group Captain. A Squadron Leader served as the Chief Flying Instructor and another as the Chief Ground School Instructor. At each SFTS there were two or more flying training squadrons with two flights each led by Flight Lieutenants. Most flying instructors were commissioned officers (Flying Officers and Pilot Officers) and a few were non-commissioned officers (Sergeants, Flight Sergeants, and Warrant Officers). Ground school instructors in navigation and armament were usually commissioned officers.

Local flying clubs contracted to operate the Elementary Flying Training Schools and private air transport companies contracted to operate the Air Observer Schools. A civilian General Manager (sometimes called an Operational Manager) was in charge of each civilian-managed station and a senior RCAF officer (a Squadron Leader or a Wing Commander) served as Chief Supervisory officer responsible for all the training. At first civilian pilots did the flying instruction at EFTSs but later, when the civilian source became exhausted, some RCAF instructors were given leave to instruct out of uniform at EFTSs. Civilian pilots flew the observers and navigators in their exercises at AOSs although many of these pilots were later given the opportunity to join the RCAF, take an SFTS course to get their wings, and then take other postings. RCAF officers gave the ground school instruction at the EFTSs and the AOSs. Navigation and Bombing and Gunnery Schools were staffed and managed by the RCAF as were the specialist schools.

The kinds of aircraft used by the RCAF at the flying schools remained largely unchanged during the life of the plan. Most elementary training schools used Tiger Moths and Fleet Finches while some used Cornells and Stearmans. Most service flying schools giving single-engine instruction used North American Harvards as trainers. (Yales were used at some schools when Harvards were in short supply at first.) Cessna Cranes and Avro Ansons were the aircraft used at the majority of twin-engine SFTSs but some of the RAF "transferred" schools used Oxfords. Ansons were used fairly exclusively for training observers and navigators at Air Observer Schools and Ansons, Fairey Battles, Bolingbrokes, and Lysanders were the aircraft used at Bombing and Gunnery Schools. When the students were "winged" and posted to Operational Training Units, they flew aircraft of the type they would be using on operations — Hurricanes, Cansos, Hudsons, Liberators, Mosquitoes, etc.

When the war began, the RCAF had but five airfields with six more under construction. Even before the BCATP was launched, the RCAF and National Defence had reached an agreement with the Department of Transport under which Transport officials would select airfield sites and, after RCAF approval, would develop them into landing fields. The RCAF undertook to design the control towers, hangars, and ground school buildings and hire civilian contractors to erect them.

INITIAL REQUIREMENTS

In November 1939, John Patterson, knowing that negotiations were underway regarding an aircrew training program in Canada, had begun to correspond with the British Meteorological Office to seek information about the meteorological requirements in the Royal Air Force flying training programs. As the weeks went by he doubtlessly also discussed the subject with the Royal Canadian Air Force and with his superiors in the Department of Transport. So, a few weeks after the agreement was signed, it would have not been a surprise to him when he was given a copy of a letter from the Department of National Defence requesting that the Department of Transport provide meteorological services at the training plan stations.

The letter of request, dated March 20, 1940, was brief. It simply asked for the provision of "Meteorological instructors that will be necessary under the British Commonwealth Air Training Plan." The letter stated that "These gentlemen will be required not only to carry out normal forecasting and reporting duties, but to lecture pupils passing through the schools on the subject of Meteorology in accordance with the syllabus attached." The letter also advised that the RCAF intended to provide one or more untrained airmen at each station to assist the meteorologist. An appendix was attached listing twenty-six training stations that would be opened and the dates when a meteorological instructor would be required at each. Further, National Defence asked that a meteorological advisor be posted to Air Force headquarters in Ottawa. The postings were to commence on April 1 and all station meteorological sections were to be manned by March 1942.[80]

While no internal memoranda or records are available revealing the considerable discussions that must have taken place at meteorological headquarters after receipt of this request, John Patterson was able to submit a plan to his department within a few days. The department agreed with the contents and, early in April 1940, the Deputy Minister of Transport responded to DND that his department was prepared to undertake the service requested and enclosed a copy of Patterson's plan. In it, Patterson recommended that a senior meteorologist, J.R.H. Noble, be seconded to RCAF headquarters in Ottawa and that another experienced meteorologist, D.B. Kennedy, be sent to the air navigation school at Trenton. Further, Patterson stated that a number of his Meteorological Assistants Grade 3 at the civil airways stations had the qualifications to meet the requirements at the training stations and could be posted there. However, he added, it would be necessary to recruit and train additional assistants for the RCAF as well as replacements for those leaving the civil stations.

In the discussions that followed it was revealed that National Defence believed that teletype communications were not required at every training school since all schools were not so equipped in Britain. But, Patterson was quick to point out, the distances between the new Canadian training schools would be much greater than the distances between schools in Britain and so teletype installations would be most desirable at each school. It was conceded, however, that when a training school was located on an airfield where there already was a civilian meteorological office with teletype connections, a school need not be so equipped.

It was agreed that the major requirement of the training program was to prepare pilots and other aircrew for operations in Britain and Europe. To assist in this regard the British Meteorological Office provided samples of manuals, forms, and other documents used in Britain so that the Canadian trainees might become aware of the meteorological organization and procedures practiced by the operational RAF. This was necessary since many of the definitions, forms, and procedures used in North America were different from those used in Britain. The London Meteorological Office offered to send a meteorologist to assist in working out the proper procedures for a military air training meteorological organization but very soon the British realized that they had an insufficient number of meteorologists to do this. Instead, the director of the Meteorological Office asked Patterson to send a Canadian meteorologist to London but Patterson claimed he could not spare one and so the instruction program was developed somewhat independently in Canada.

Patterson's first personal contact with an RAF meteorologist came in August 1940 when Group Captain Entwistle, who was seconded to the RAF from the Meteorological Office, visited Canada and attended a meeting at Toronto of RCAF and Meteorological Division training officers. By this time the Meteorological Division had prepared three sets of lecture notes (for the SFTSs, AOSs, and Central Flying School) and these were reviewed and accepted. Since no suitable textbook was yet available it was recommended that copies of the notes be mimeographed and sent to the schools so each student could have a copy.

The meeting agreed that it was essential that each school be connected to the meteorological teletype system regardless of the cost. (Costs were then estimated to be several thousand dollars for each school.) This would allow the plotting and analysis of daily weather maps which student aircrew could inspect in order to co-ordinate their actual experiences in the air with the maps and reports. There were discussions regarding the most suitable official title to use for the men who were even then being unofficially called "metmen" but no conclusion was reached.[81]

STRUCTURE OF THE PLAN

On April 29, 1940, the first group of BCATP aircrew trainees left the Manning Depot and arrived at Initial Training School in Toronto. Here, over a four week period, they were given lectures in navigation, aerodynamics, and armament. They took physical training and were sorted into different categories for further training. Those considered to have high enough physical and educational qualifications to master the pilot or observer courses were selected for either one while the less qualified of those who had aircrew potential were sent for training to become wireless operators and air gunners.

For student pilots the next step was to an Elementary Flying Training School (EFTS) operated for the RCAF by a local flying club. (There were also EFTSs in other commonwealth countries but all advanced flying training was carried out in Canada.) Civilian pilots gave the flying instruction at the EFTSs and RCAF officers the ground school instruction. The course took eight weeks during which time each student pilot took dual instruction and was expected to be ready for his first solo flight. (The length of the flying training courses varied somewhat over the course of the war according to the need for operational aircrew.)

At first it was considered that there was no need for metmen at EFTSs. Weather forecasts were obtained by telephone from a nearby civil meteorological office or from the nearest RCAF school staffed with metmen or meteorologists. However, as more metmen became available during the last year or so of the war, metmen were posted to a few EFTSs to meet special needs.

EFTS graduates were then posted to a Service Flying Training School (SFTS). Although navigation instructors gave some lectures in meteorology to the trainees at EFTSs it was at an SFTS that the student was first obliged to study and pass a course in meteorology given by a metman. A new group of thirty or so student pilots arrived at each SFTS station every twenty-four days and graduated with their wings in about three months. Only ten hours of instruction in meteorology were scheduled at first but, by 1944, this was increased to sixteen hours. The courses and lectures were so arranged that there were never more than three courses taking ground school instruction at the same time and each course had one specific metman as instructor. The metmen also analyzed weather maps, issued forecasts, and gave weather briefings to students and flying instructors before flying exercises.

Those selected to be navigators and air bombers were sent to an Air Observer School (AOS) after Initial Training School. Early in the war, the AOS graduates were classified as "air observers," but, early in 1942, when it was determined that a second pilot was not necessary on heavy bombers, the trade of "air bombers" was introduced and the observers became "navigators" with training in electronic equipment added to their courses.

The AOSs were operated for the RCAF by private air transport firms. In general the staff pilots who flew the students on their exercises were civilians while RCAF officers gave the ground school instruction except for meteorology. At first, the AOS meteorological course took fourteen weeks but by September 1942, when the course had been divided, the navigators' course had been extended to twenty weeks while the air bomber course took six and later ten weeks. Each AOS required a larger staff of metmen than did an SFTS because of the longer and more intensive course in meteorology at these schools.

The students at Bombing and Gunnery Schools (B&GSs) were air bombers, wireless operators, air gunners, and observers on courses of varying lengths. There was no meteorological instruction given to these aircrew students but by 1944, when metmen became more numerous, some were posted to several schools of this type for forecasting duties as longer training flights were introduced into the program.

At the Operational Training Units (OTUs) graduate pilots, navigators, air bombers, and wireless air gunners were "crewed up" into teams and began flying exercises to duplicate operational flights. Both meteorologists and metmen were posted to OTUs to prepare forecasts and brief the crews before their flights and to give lectures in advanced aviation meteorology. Other schools to which both meteorologists and metmen were posted were Air Navigation Schools (ANSs), General Reconnaissance Schools (GRSs), and specialist schools for flying instructors, navigation, and instrument flying, as well as to Trenton where the Central Flying School and the Central Navigation School were first located.

Most metmen serving at BCATP schools were at Service Flying Training Schools and Air Observer Schools. For example, in March 1944, just before the training plan agreement ended, there were ninety metmen at twenty-seven SFTSs and forty-two metmen at eight AOSs. By that time the establishment at most SFTSs called for three metmen and for five or six at AOSs because of the heavy lecturing load at the latter.

DUTIES AND RESPONSIBILITIES

The duties and responsibilities of the metmen (meteorological officers) at the air training schools were outlined in 1940 by John Patterson. In general, these applied to all types of schools and remained valid over the life of the BCATP. In summary these were:

1. Supervise the activities of the meteorological section at each station and be in charge of airmen assistants assigned for meteorological duties;
2. Prepare at least one weather map each day and provide forecasts and weather information as required for flying operations, brief the pilots and students before their flights and keep air navigation personnel informed of the weather in the region where flying is taking place;
3. Give a course of lectures to each class of students in accordance with a prescribed syllabus;
4. Give advice and explanation of the weather to students in connection with their course and discuss with them, after their flights, the forecasts and weather conditions expected and experienced;
5. Advise the RCAF staff at all times regarding weather problems;
6. Notify the station commander of any changes in the program and also of any absences from the station.

During his visit to meteorological headquarters in August 1940, RAF Group Captain Entwistle made clear the necessity for meteorological training in the BCATP:

> In view of the dependence of modern air operations performing bombings and reconnaissance flights on accurate meteorological information, it cannot be emphasized too strongly that air pilots and observers during their training should acquire the habit of consulting the meteorological office before undertaking flights, and also by discussing the weather conditions during the flight with the meteorologists after landing. It is only in this way that pilots and navigators will become familiar with the practical application of meteorology to flying, which forms the subject of lectures given during the course of training. In other words, the meteorological forecasting service on the airfield is just as important a part of meteorological instruction as the actual lectures. [Quoted by Patterson in his unpublished manuscript "History of the Meteorological Service of Canada During World War II."]

INCREASED REQUIREMENTS

Air training under the BCATP was just getting underway in the spring of 1940 when France fell to the enemy. Cabinet level discussions took place in Britain as to whether or not the plan should be dropped and all qualified aircrew in Canada, including instructors, be sent overseas to bolster the RAF in their fight against the expected Nazi invasion. Fortunately, a long-range view prevailed and the plan was left intact but the training programs were accelerated for awhile to shorten the period the students were on course.

A few weeks after the training schools began to open the RCAF found that Patterson's warnings that one metman could not possibly handle all the duties required of him each day was right. This allowed Patterson to increase his estimate of need to twenty-five rather than the originally estimated fifteen new metmen and, at the end of August, he increased this number to sixty-five. (The number kept increasing throughout the life of the plan, as the value of meteorology became more and more apparent to the air force.) To cover the initial shortage of metmen, until Canadians could be recruited and trained, Patterson sought and obtained authority to ask the United Kingdom Air Ministry for the loan of twelve meteorological officers and the RCAF agreed to be financially responsible for their pay and upkeep in Canada.

Although the British were sympathetic at first, it quickly became evident that their own demands for trained meteorological officers had increased so rapidly that there were none available to send to Canada. Accordingly, during the summer of 1940, the question of recruiting and training additional meteorological staff in Canada was a problem of major proportions for Patterson and his staff. It was decided that the graduate course in meteorology at the University of Toronto would again be given in the coming academic year but it should be shortened by a month to make the new meteorologists available sooner. But this would produce only a dozen or so new meteorologists while scores of men with somewhat less training were required almost immediately at the RCAF training stations. So it was decided to institute a short intensive course, given within the Meteorological Division, to train men to carry out the required duties in the purely war positions at the schools and at Home War units. The story of the recruiting and training of these metmen (Meteorological Assistants Grade 3) and their association with the RCAF is given in an earlier chapter.

The first Intensive Course in Meteorology with nineteen mathematics and physics graduates began in the fall of 1940 and these men became available for posting as metmen to the air schools in early March 1941. Over the next three years, on several occasions, the RCAF increased the establishment for metmen at the training stations and made repeated requests for the positions to be filled. To meet this increasing need the Meteorological Division recruited, hired, and trained 373 metmen in twelve courses between late 1940 and the summer of 1944.

By the end of 1940, in addition to the need for metmen who were trained to become "dependent" forecasters, the Meteorological Division needed additional meteorologists to become "independent" forecasters as well. The RCAF requested the more qualified meteorologists rather than metmen for some operational stations and for the Operational Training Units it was planning to open. Also, the Meteorological Division needed more meteorologists to enable it to open additional civil aviation forecast offices. There were seven meteorologists graduating and available in the spring of 1941 but many more were needed. Accordingly it was planned to give an advanced course in meteorology to twenty or so of well-qualified metmen and the first such course was held from October 1941 to January 1942. The course was successful and seven more were given over the next four years in which about 100 metmen were trained to become meteorologists and to do independent forecasting.

PIONEER OFFICES

Beginning in the spring of 1940, staffing the meteorological sections at the new BCATP schools became a continuing and major problem for John Patterson and his Head Office staff. As soon as he arrived in Ottawa as the meteorological advisor at RCAF headquarters in April, Reg Noble became the contact between the RCAF and Patterson regarding the RCAF requirements at the new schools. In addition, he was thrust into the preparation of lecture notes for use by the metmen at the schools. Here, Des Kennedy, who had been posted to the RCAF's premier flying training base in May, was of great assistance to Noble. In the summer of 1940 they shaped the syllabi for the SFTS and AOS courses and in the fall Kennedy was brought back to meteorological headquarters in Toronto to instruct student-metmen and to continue work on preparing lecture notes for student aircrew in the various courses.

Fortunately, amongst the observing staff hired earlier for the terminal and intermediate airports along the transcontinental airway, there were a number of well-qualified university graduates who were happy to take such technical jobs in the Depression. By 1940 several had become very interested in meteorology and had acquired considerable knowledge of the subject. This was noted by their supervisors and it was from this group of men, then classified as Meteorological Assistants Grades 2 or 3, that Patterson would draw his first men for posting to the RCAF training schools as metmen. Also, a few individuals, who had failed to meet the high academic standards demanded from students in the Toronto MA course in physics (meteorology), had not been hired as meteorologists but had taken employment as observers. These men also became a source for metman postings.

Both the Central Flying School and an Air Navigation School existed at Trenton when Kennedy was posted there in May 1940. The meteorological section consisted of five airmen and was headed by an RCAF Warrant Officer. The staff obtained synoptic observations from Stirling by motor transport each day to enable them to plot weather maps while forecasts and some hourly observations were received by telephone. (Stirling, some seventeen miles distant, was a radio range station on the airway and so was connected to the airways meteorological teletype circuit.) Kennedy was primarily occupied with briefing and preparing lectures for different types of navigation classes and was also involved in preparing syllabi and lecture notes for other types of schools. When he was moved to Toronto in October his replacement was Jim Leaver.

There was an RCAF flying training school at Camp Borden prior to the April 1940 launch of the BCATP when the school was absorbed into the plan as No. 1 Service Flying Training School. Bill Green, with observing and some pilot briefing experience at the North Bay intermediate airport meteorological office was posted to the school in April. Weather maps were sent by mail from Toronto to Camp Borden for the first few weeks and it was not until July that the RCAF authorized trial installations of teletypes at Camp Borden and Trenton. Meanwhile, Green, guided by correspondence with Noble, prepared notes for the lectures to the student pilots in the weeks before Toronto headquarters was able to develop a standard set of notes for use at all SFTSs.

The Camp Borden meteorological section was in many respects the pioneer office where difficulties and insufficient resources first became apparent. When Fred Turnbull, officer-in-charge of the Toronto Malton District Aviation Forecast Office, visited the station in December 1940, he reported that the meteorological office was remotely located, there was not sufficient classroom space, the length of the course had been severely cut, and that one metman assisted by one airman assistant could not possibly do the work required.

It was somewhat different when the first Air Observer School opened at Toronto Malton in May 1940. The DAFO at the airport was then operating with a full-time staff under Turnbull (for several months before the war the DAFO had been staffed by assignment from the available meteorologists at Toronto headquarters). Turnbull assumed responsibility for lecturing the student observers and briefing the civilian pilots before flights. In this observers Fred Patterson and G.K. Boyd assisted him. Turnbull continued to look after the AOS requirements at Malton until a graduate of the first course for metmen (Edward Longhurst) was sent there in March 1941.

STAFFING

When the second SFTS opened at Ottawa Uplands in August 1940, Don Perrie, an observer with experience at the Toronto Malton and Ottawa airports, was sent there. Meanwhile, a few other college graduate observers were considered experienced and capable enough to open meteorological offices at more new schools and they were sent either to a regional DAFO and/or to the Camp Borden SFTS to become acquainted with interpreting forecasts and providing pilot briefings for the air force. Fred Patterson spent four months with Green at Camp Borden before going in October to start a meteorological section at Kingston No. 31 SFTS (this was a RAF "transferred school" where British meteorologists were to arrive a few months later). Three observers from western Canada were brought east that summer for two or three weeks to learn what would be expected of them at the new schools. Harlan Thompson from Lethbridge and Dave Strachan from Vancouver went to Camp Borden and George Robertson from Edmonton went to Toronto Malton. They then returned to the West to organize meteorological sections at new schools.

In September, Elmer Stevens who had observing experience at Halifax, opened the Calgary SFTS meteorological office and Strachan opened one at the Saskatoon SFTS. C.B. McIntyre, an observer at Halifax, went to open one at the Moncton SFTS and Harlan Thompson the one at Claresholm, Alberta. Other SFTSs opening in the fall of 1940 were in Ontario where Wendell Smith, a graduate physicist who had just joined the Meteorological Division, opened an office at Dunnville and Peter Sandiford, another well-qualified new employee, opened the Brantford office. C.B. McIntyre opened another at the Summerside, PEI, SFTS in January 1941. Norm Powe, another experienced observer, replaced him at Moncton. In all, nine Service Flying Training Schools opened and meteorological sections organized before trained metmen became available at the end of February 1941.

That Patterson and the Meteorological Division were able to accomplish this to the satisfaction of the RCAF is remarkable.

The first Air Observer School opened at Toronto Malton airport in May 1940. As already noted, Fred Turnbull, officer-in-charge of the Malton DAFO, undertook to arrange for flying forecasts and ground school lectures for the courses until the spring of 1941 when a trained metman was posted there. In August 1940 the second AOS opened at the Edmonton airport where Tom How was the meteorologist in charge of a small aviation meteorological office. After his two weeks or so at Malton, George Robertson returned to Edmonton to lecture in the new school where he was joined by Fred Kelly and then Syd Buckler, a meteorologist, because of the long range flying from the school. By the end of 1940 other AOSs opened at Regina where Vic Beirnes, an experienced Winnipeg observer, and Elmer Stevens from the Calgary SFTS opened the meteorological office and at London where Jack Wingfield, a Toronto Malton observer, was put in charge of the new meteorological section. Early in January 1941 the fifth AOS opened at the Winnipeg airport with Jack Labelle, who had been observing there, in charge of the meteorological section.

GROUND SCHOOL

Another major problem for Patterson and the meteorologists involved with the BCATP at the beginning was the preparation of meteorological syllabi, notes, and examinations for the metmen to use in the ground schools at the service and air observer schools. Copies of such material used by the Meteorological Office in peacetime RAF training in Britain were sent to Canada and these were used as models on which adjustments were made to conform with the length of the courses and other factors. Lecture notes were prepared one by one, often just in time for the next lecture. Reg Noble was the key meteorologist in preparing the syllabi and course notes working by mail with Turnbull at Malton and Green at Camp Borden who were already giving the first courses. Dr. Bernhard Haurwitz, the ex-patriot German physicist who was at the Toronto Head Office those years, was the final editor to ensure the texts were clear and unambiguous. In October 1940, Des Kennedy was brought to Head Office to work on the training material and to instruct the first intensive course for metmen. Jim Leaver replaced him at Trenton. Before the end of the year, Leaver was also brought to Toronto to lecture and supervise the metmen courses.

Without a text it was necessary to get the draft lectures mimeographed and out to the schools one or two at a time before each was required on the courses. The AOS course was longer that that at the SFTSs and so the notes for those courses were more extensive. It would be 1941 before the first textbooks were published by the Meteorological Division and made available for student use — *Elementary Meteorology for Service Flying Training Schools* and *Meteorology for Pilots and Navigators*. These were in short supply at first and bureaucracy was such that the students were asked to purchase their copies for a few cents each if they wished to keep their copy after graduation!

FACILITIES AND EQUIPMENT

At the Service Flying Training Schools the meteorological office was usually located on the second or third floor of the control tower. The first floor of the tower building accommodated the Commanding Officer, the station adjutant, and other administrative staff. Above there were usually two small one-roomed floors topped by a glass-enclosed "tower" room, where the flight controllers worked. The "Met Office" and a radio communications room were below the tower room with an open stairway through each. At the Air Observer Schools and other training schools the meteorological section was more likely to be housed in the end of a hangar or in the ground school building.

The typical SFTS Meteorological Office had four windows, arranged in bay window style on the flying field side, which allowed an excellent view of the field including the runways, the taxi strips, and the aircraft parking areas. The flight controllers, in their glassed-in tower room, were often visited by the metman when he was worried about his local weather forecast and wanted to check on the approaching weather.

The meteorological office was separated from the stairway by a counter that could be used when giving informal briefings to students, instructors, and air traffic control officers. There was usually one map table (a wooden sloping table with a glass top) in the centre of the room with a phone mounted nearby. The teletype, a barometer, a typewriter, and one or two tables took up the remainder of the outside wall space. There was also a swivelled stand or some such device for holding teletype printouts (observations and forecasts) for ready viewing. A large frame on which to hang two or three weather maps usually occupied one wall above a table. Although not spacious, there was ample working room in the office for a metman and one or two assistants.

Both the Meteorological Division and the RCAF provided the equipment for work in the meteorological office. The provision of suitable space for the office was the responsibility of the RCAF along with the furniture, a typewriter, and a telephone. The Meteorological Division was responsible for the teletype and supplied the meteorological observing instruments — thermometers, barometer, barograph, thermograph, an anemometer, and a Stevenson screen to house the thermometers. Gelatinous "Ditto" machines for duplicating forecasts were provided by the Meteorological Division along with all the paper (forms, maps, texts, etc.) used in the office and in ground school. The Transport Department provided ceiling projectors for use in measuring the height of cloud bases.

There was, of course, no need for instruments at those schools collocated at airports where there already was a civil weather reporting station. The first Air Observer Schools were so located and it was determined initially that the meteorological offices at those schools did not need a teletype connection, but for convenience and efficiency in the training program this policy was soon changed. (All expenses incurred by the Department of Transport and the Meteorological Division for service to the RCAF and the BCATP was covered each year in special war service allocations.)

AIRMEN ASSISTANTS

Mention has already been made of a course conducted by the air force in 1939–1940 for airmen when it was thought that high school graduates could be trained to be meteorologists. By the time the course was completed, agreement had been reached with the Meteorological Division for the secondment of professionals — meteorologists and metmen — so the RCAF posted most of their "meteorologists" to operational bases on the coasts. Some of these airmen became non-commissioned officers and handled the work in operational meteorological sections for a year or more until meteorologists and metmen became available. They were an outstanding group of technicians but few remained in meteorology because of the lack of opportunity for advancement once the professional civilians arrived.

One airman assistant was promised for each meteorological section to assist the civilian metman. However, the first airmen assigned for meteorological work were usually the lowest grade of aircraftmen (AC2s). They often had neither the aptitude nor desire to work in a meteorological section. Soon the airmen did become Leading Aircraftmen (LACs) but, on some stations, they were subject to parades and other station obligations which interfered with their scheduled work times in the office. The best of them, since they were General Duties airmen, asked for postings out of the meteorological section to some other area where they would be classified as having a "trade."

Except for those who had taken the 1939 course, there was no organized meteorological training in the RCAF for the airmen assistants during the first two or three years. Some airmen were sent to a District Aviation Forecast Office for a few weeks but most were sent directly to a training station where the metman was responsible for their training. Very soon the RCAF realized that two or three airmen were required in each meteorological section and when this happened the senior airman trained new assistants on the job. At least one airman was on duty whenever flying was in progress although it was difficult to cover both day and night flying with only two or three airmen on staff.

On shift, the airman's time was taken by map plotting, taking a weather observation each hour, monitoring the teletype, typing the forecasts written by the metman, and distributing them to the various flight and administrative offices.[82] By early 1941 one airman had been promoted to the non-commissioned officer rank (NCO) of corporal at most training schools.

By 1942, and after strong Meteorological Division appeals to the Air Force, a Meteorological Observer trade was established for which an airman could be examined and qualify. This gave him more pay and allowed him to wear a shoulder flash identifying his trade.

TRANSFERRED SCHOOLS

Independently from the British Commonwealth Air Training Plan, the Royal Air Force continued its training program during the early months of the war. With the loss of the Low Countries and the fall of France to the enemy, the operational pressure on airfields and airspace made it difficult if not impossible to continue training activities in Britain. Accordingly, in July 1940, the British government requested permission of the Canadian government to transfer four Service Flying Training Schools to this country. Canada agreed and volunteered to take additional schools if Britain found it necessary to move them. Within days, this offer was accepted; the RAF asked and consent was given to send more than a dozen training schools to Canada.[83]

The first school to leave Britain for Canada was an SFTS that arrived in Kingston, Ontario, early in September 1940. It had been planned that the schools would move complete with aircraft, equipment, and staff but no meteorological personnel came with the school that was designated No. 31 SFTS. (All RAF transferred schools were given numbers 31 and above to differentiate them from the BCATP schools whose numbers started at 1.) To help the school begin its training program the Meteorological Division sent Fred Patterson, an observer who had recently spent four months at the new Camp Borden SFTS, to the school. The Kingston RAF meteorological contingent arrived in March 1941 and Patterson moved on to western Canada where, over the next year, he opened eight more meteorological sections at transferred schools. His stay at each varied from a week to more than three months since he remained at each school until a Canadian metman became available for posting there.

By the end of 1940, RAF meteorological staff had come with the navigation schools that arrived at Port Albert and Hamilton Mount Hope in Ontario and at Charlottetown, PEI.[84] But the RAF made so many demands for meteorological staff in Britain and in other operational theatres that the Air Ministry had to advise Canada that no more could be sent with the transferred schools. As a result of negotiations involving the Air Ministry, the RCAF, and the Meteorological Division it was agreed that Canadian metmen would be provided to these schools even if this did slow down the proper staffing of the BCATP schools. The Meteorological Division also supplied the meteorological instruments, weather maps, manuals, and instruction books and arranged for teletype connections. Later on, in 1943, a dozen or so RAF meteorological personnel were invited to participate in the Short Intensive Courses where metmen were trained.

The provision of airfields and buildings for the transferred schools proved to be a major challenge for the Department of Transport. Some sites were chosen that had been under development as emergency landing fields along the transcontinental airway and others had been originally designated as "relief fields" for specific training stations. Already working with civilian contractors at top speed to ready facilities at many sites for the BCATP schools, Transport and the RCAF managed to have basic facilities ready for the RAF schools when they arrived although some stations were by no means completed by that time.

The schools continued to arrive throughout 1941 and by the summer of 1942 nineteen RAF transferred schools were training aircrew in Canada in close co-operation with the RCAF. Canadian metmen staffed the meteorological sections at fifteen of these schools and British personnel at the original four schools. The courses at the schools were about the same as those given at the BCATP schools and so there was no difficulty in moving metmen around if necessary. (In time, following negotiations with the Air Ministry and the countries that signed the BCATP agreement, the transferred schools became part of the British Commonwealth Air Training Plan.)

[77] Introduction by W.A.B. Douglas in F.J. Hatch, *Aerodrome of Democracy: Canada and the British Commonwealth Air Training Plan 1939–1945* (Ottawa: Department of National Defence, 1983) p. xvii.

[78] F.J. Hatch, *Aerodrome of Democracy: Canada and the British Commonwealth Air Training Plan 1939–1945*; W.A.B. Douglas, *The Creation of a National Air Force* (Toronto: University of Toronto Press, 1986), pp. 154–293; James N. Williams, *The*

Plan: Memories of the BCATP (Stittsville: Canada's Wings, 1984); Peter C. Conrad, *Training for Victory: The British Commonwealth Air Training Plan in the West* (Saskatoon: Western Producer Prairie Books, 1989), Spencer Dunmore, *Wings for Victory* (Toronto: McClelland and Stewart, 1994).

[79] Hatch, *Aerodrome of Democracy*, pp. 1–26.

[80] AES, Archives, Acting Deputy Minister of National Defence to Deputy Minister of Transport, 20 March 1940.

[81] AES, Archives, File 5930-1-0.

[82] Letter from Paige Knight, a wartime RCAF observer who became a meteorologist after the war, September 1991.

[83] Hatch, *Aerodrome*, pp. 62–83.

[84] Don McNaughton, a draft manuscript "United Kingdom Meteorologists in Canada during World War II" sent to the author in 1991.

CHAPTER 14

TRAINING AT ITS PEAK

The first year of the BCATP had been hectic for John Patterson and his senior meteorologists. Construction of the airfields and the training school buildings proceeded faster than planned allowing some schools to open four to six weeks ahead of schedule in 1940 and as many as thirty-two weeks ahead in 1942. With the fall of France and the expected invasion of Britain there was a move towards abandoning the Plan and sending all trained aircrew overseas. Fortunately this did not happen but the pilot training course was accelerated markedly. The EFTS course was shortened by a week in mid-1940 and the SFTS reduced from sixteen to fourteen weeks in July and to ten weeks in August. There were only minor changes made in the other aircrew training programs but the acceleration of pilot training made the staffing problem more acute for Patterson. (The pilot courses were lengthened again within a few months to improve the quality of the graduates.)

Further, the RCAF had been slow to realize the need for teletype communications and the need for more than one metman and one or two airmen assistants at each station. Both situations were rectified by the end of 1940 but Patterson found that "planning and execution had become inextricably mixed and one plan had hardly been made and begun to be put into effect before further experience resulted in alterations to the plan."[85]

THE AGREEMENT RENEWED

The original BCATP agreement was due to expire on March 31, 1943. By 1941 the training scheme had far outgrown the terms of the original agreement and the costs to Canada had been greatly increased. So the Canadian government began to press the British to renegotiate the terms of the plan. This was accomplished at an Ottawa conference in May 1942 which had grown to become a general conference on air training, which included the United States and other allies. The general conference had no control over the BCATP but did provide a valuable link towards developing air training on a nearly common basis in Canada and the United States.

The second part of the conference included only the four commonwealth countries that had first agreed in December 1939 — United Kingdom, Australia, New Zealand, and Canada. All parties viewed

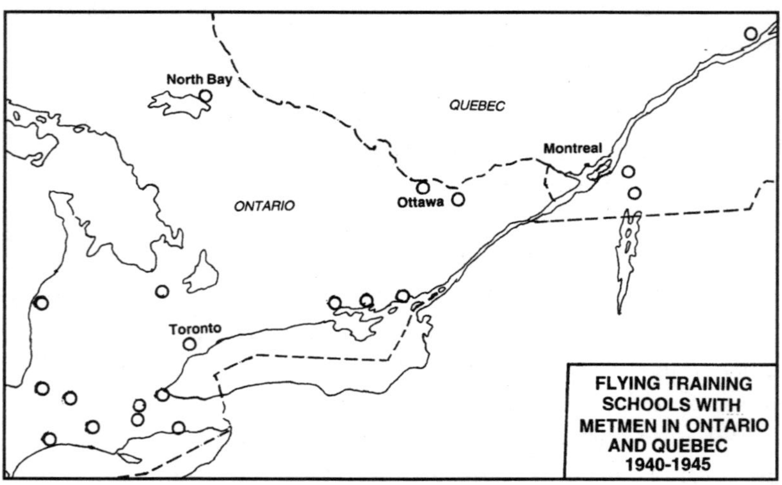

FIGURE 6

Flying training schools in Ontario and Quebec where metmen were stationed in wartime.

the training plan as a success and agreed that Canada was doing well at operating it. The British agreed that the transferred RAF schools should be completely amalgamated with the original schools in the plan while recognizing British command at those schools. Various financial and control aspects of the plan were discussed and settled sufficiently to allow training to proceed under the new agreement.

SERVICE FLYING TRAINING SCHOOLS

During 1941 the new SFTSs continued to open ahead of schedule and in each case the Meteorological Division was successful in posting a metman on time to open a meteorological section and begin the necessary briefing and lecturing programs. It was March 1941 before metmen from the first Intensive Course in Meteorology became available for posting and the latter part of that year before it was possible to post a second metman to most SFTSs. Early that year, when it became evident that no professional meteorological staff would accompany nine more RAF transferred SFTSs to Canada, the need for metmen intensified.

This need continued to outstrip the supply from the first course or two and more experienced observers were sent to open stations in 1941. Don McGeary, after considerable observing experience at Regina, was sent to the new Rivers Air Navigation School for orientation and then to open a new meteorological section at the Dauphin SFTS. Hugh Cameron, also an observer at Regina, was sent to Claresholm to open another section without the background of a training course. However, as course graduates began to become available every three to four months later in 1941, John Patterson sent them to open sections at new SFTSs, even if they had only a few weeks experience at another RCAF station or at a civil forecast office. Before the end of 1941, additional SFTSs had been opened at Yorkton by R. Pattison, at Brandon by Alex MacVicar, at Montreal St. Hubert by Merv Fleming, and in Ontario at Aylmer by Al Mason and at Hagersville by Wilbur Sly. Later, in 1943, three more RCAF SFTSs were opened at Souris and Gimli in Manitoba and at Vulcan, Alberta; by that time there were enough trained metmen to staff the new sections.

But, in 1941, there were just not a sufficient number of capable experienced observers or new metmen to provide a trained metman to each RAF SFTS when it opened. The schools began arriving in December 1940 and it would be several weeks or even months before a Canadian metman arrived. In nine of the ten transferred schools that metman was Fred Patterson. A Toronto Malton observer with brief experience at the Malton AOS and the Camp Borden SFTS before going to Kingston to set up a meteorological section for the first RAF SFTS to come to Canada, Patterson then spent the next eighteen months opening sections at the RAF stations. He spent four months at Carberry in 1941 and then a few weeks at each of Moose Jaw, Medicine Hat, and North Battleford. Later in 1941 he opened schools at Penhold, Swift Current, Weyburn, and Calgary (Currie Field). In each instance another metman arrived before or soon after he left. Only the RAF SFTS at Estevan, which did not open until April 1942, missed having its meteorological section opened by Patterson. (After his extensive western road trip Patterson was posted to personnel work at the Toronto headquarters and apparently did no more travelling.)

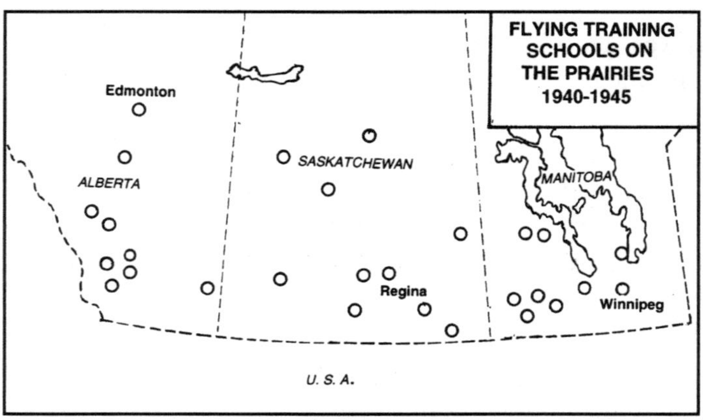

FIGURE 7

Flying training schools in Alberta, Saskatchewan, and Manitoba where metmen were stationed in wartime.

There was a common instruction syllabus at the Service Flying Training Schools and consequently the requirement for meteorological work was almost identical at the various schools. The number of lecture hours allowed for meteorology in ground school varied a little according to the meteorological-awareness of the Chief Ground School Instructor. Similarly the attention and time given to weather briefing before training flights depended to some extent on the weather-consciousness of the Chief Flying Instructor. However, as younger and more recently trained senior officers became responsible for the training program the importance of meteorology was increasingly recognized and more time allotted to instruction and briefings.

Of course, the attitude of the metmen towards their job and their interest and enthusiasm for meteorology certainly varied from man to man and station to station. But, overall, station and personnel inspection reports by the meteorologists on the Visiting Flights were remarkably good. This was probably because most metmen, although they had previously no knowledge or interest in meteorology, became interested in the subject and enjoyed the work. It is worth noting that about one third of the wartime metmen abandoned earlier career plans for the postwar period and decided to make meteorology their life work.

Part of Chapter 15 is devoted to describing activities at a representative Service Flying Training School — No. 12 SFTS, Brandon, Manitoba.

AIR OBSERVER SCHOOLS

There were fewer Air Observer Schools in the British Commonwealth Air Training Plan than Service Flying Training Schools but more metmen were required at the AOSs because of the heavy lecturing program. A variety of aircrew were trained at the AOSs compared to the SFTSs where only pilots were trained. At first, all AOS students were trained to become air observers, a trade that included bomb aiming and air gunnery as well as navigating.

Then, a second pilot was dropped from operational heavy bomber aircrew requirements and more attention was given to navigation, bombing, and gunnery instruction for the other crew members. The new training requirements meant an increase in size of most Air Observer Schools from a capacity of 126 to 676 trainees. This meant that more metmen would be required at AOSs to give instruction in meteorology and more briefings before training flights.

New courses were introduced to produce air navigators, air bombers, air gunners, and wireless operators in courses of varying lengths. Most aircrew students at AOSs took additional courses at Bombing and Gunnery Schools and Air Navigation Schools before getting their wings.

Four AOSs had been opened in 1940 and six more followed in 1941 — Winnipeg; Prince Albert, Saskatchewan; Portage la Prairie, Manitoba; Ancienne Lorette, Quebec; St. Jean, Quebec; and Chatham, New Brunswick. No more were required as several schools became "double" schools handling twice the number of students they had trained at the beginning. The length of the original AOS course was not notably shortened during the 1940–1941 period of training acceleration but, in September 1941, the course length for observers was increased from twelve to fourteen weeks and a class of twenty-six entered every two weeks instead of every four weeks.

As at Toronto Malton, the Winnipeg AOS was collocated at the airport with a District Aviation Forecast Office which meant that the airmen assistants did not take weather observations and no teletype was installed at the school for the first eighteen months. Aircrew briefing was done from forecasts and maps provided by the DAFO. Jack Labelle, an experienced observer at Winnipeg, opened the meteorological section at the AOS in January 1941. The Prince Albert AOS opened in March with Elmer Stevens responsible for the meteorological section. Stevens, another experienced observer, had opened the Calgary SFTS several months earlier. When the next schools were opened at Portage la Prairie and Ancienne Lorette, the meteorological sections were headed by Don Currie and Ronald Glen, graduates of the first intensive course, who had each spent a month or two at other schools. Later in 1941, two experienced observers, Norm Jefferson and Conn Sutherland, headed meteorological sections at two new Air Observer Schools at St. Jean and Chatham.

Because the courses were longer and covered more meteorology at the AOSs than at the SFTSs, a second metman was badly needed from the beginning at each AOS. These were added and then, at some AOSs, the intake of students was doubled which meant that a third metman was required. But, it would be late 1942 before these AOS meteorological sections obtained a third metman.

A somewhat detailed description of activities at one AOS, No. 1 Air Observer School, Malton, Ontario, is given in chapter 15.

ELEMENTARY FLYING TRAINING SCHOOLS

There were twenty-four Elementary Flying Training Schools (EFTSs) in the original BCATP and the RAF brought six more to Canada. The original twenty-four schools were civilian operated for the RCAF and soon after they arrived in Canada the RAF schools were taken over by local flying clubs.

For two years there were no meteorological sections at EFTSs. The flying training exercises were local and of too short duration to warrant the expense of a meteorological staff. Flying forecasts were obtained from a conveniently close RCAF station staffed with metmen and, in ground school, a navigation instructor gave a short course in meteorology.

Several EFTSs were collocated at airports with Air Observer Schools. When the Prince Albert AOS closed in September 1942 it was decided to leave a meteorological section at that airport to serve itinerant RCAF aircraft as well as the EFTS. Somewhat the same situation occurred at Yorkton, Saskatchewan, where an SFTS closed in December 1944 and an EFTS moved to that station. The Yorkton meteorological section remained open for several months as another one did at St. Jean, Quebec, after the AOS was closed at that airport.

AIR NAVIGATION SCHOOLS

There had been an RCAF Air Navigation School at Trenton before the war and it was absorbed into the BCATP in April 1940 and then moved to Rivers, Manitoba, in November of that year. In the summer of 1941, a second ANS was opened at Pennfield Ridge, New Brunswick.

Because of poor flying conditions this ANS lasted less than a year and later was reopened at Charlottetown in February 1944. Three RAF transferred Air Navigation Schools came to Canada — Port Albert, Ontario, in November 1940; Hamilton's Mount Hope airport in June 1941; and Charlottetown in August 1941. British meteorological personnel accompanied these schools in their move to Canada.

Students in most aircrew courses at Air Navigation Schools had already graduated and received their wings at an SFTS or an AOS. Consequently, their ANS course in meteorology was more advanced than that given at the other schools. Their flying exercises were usually of longer distances and duration than those flown from the other schools and this required an extensive briefing program.

The Trenton/Rivers school was considered to be the centre of the entire navigation training program and its chief function was to maintain a uniformly high standard at all types of schools where navigation was taught. An extensive account of the Rivers Central Navigation School is given in the next chapter.

Ade Lenahan was the metman at the Pennfield Ridge ANS over its nine-month existence. The Royal Air Force ANS at Charlottetown was closed in September 1942 and, in 1944 when another ANS was opened at Charlottetown, A.A. Hoover was in charge of the meteorological section. Over the year and a half of operations at that school eleven different metmen served there.

GENERAL RECONNAISSANCE SCHOOLS

In addition to the RAF General Reconnaissance School (GRS) at Charlottetown, Prince Edward Island, which operated from 1941 until early 1944, the RCAF operated another GRS at Summerside. This school opened in July 1942 and closed early in February 1945. The purpose of the GRSs was to allow pilots and navigators to gain experience in flying long distances over the ocean to prepare them for posting to coastal operations squadrons.

Considerable operational flying was done from both GRS stations and paragraphs have been written about them in Chapter 8 which deals with staffing the meteorological sections at Eastern Air Command bases.

BOMBING AND GUNNERY SCHOOLS

The first Bombing and Gunnery School in the BCATP was opened at Jarvis, Ontario, in August 1940. By the end of 1941, nine such schools were training aircrew in the Prairie provinces, Ontario, and Quebec. Also, an RAF transferred B&G School came to Ontario in late 1941, and in 1943 a tenth school opened.

At first the B&G schools were used primarily to give air observers bombing instruction in ground school and practical exercises in aircraft flown by staff pilots. These students came from AOSs for a period during their observer training. Some gunnery exercises were carried out but early air gunners received most of their training after they arrived in the United Kingdom. Because the flying was local and done by experienced RCAF pilots it was considered that there was no need for a meteorological section at the B&G schools. Staff at these schools regularly telephoned the nearest RCAF station with a meteorological office for regional weather reports and forecasts. An exception was the Mont Joli Bombing and Gunnery School where metmen were stationed since an RCAF anti-submarine operational squadron was based there during the summer shipping season.

The need for meteorological services changed at most B&G schools in the spring of 1944 when navigation exercises were initiated which required weather forecasts and a briefing program. To meet this requirement, the RCAF set up an establishment of one metman and two airmen at each school. As it turned out meteorological sections were established at only four of the schools. Metmen were posted to schools at Mossbank and Dafoe, Saskatchewan, and at Fingal, Ontario, for the three to five months before the stations closed in the winter of 1944–45. The B&G School at Paulson, Manitoba, remained open for several more months in 1945 to service itinerant aircraft. When the different B&G meteorological sections were closed there were five metmen at Paulson, four at Mont Joli, and one at each of the others.

Mention has already been made in Chapter 8 of a Naval Gunners School at Yarmouth, Nova Scotia. Yarmouth was a major RCAF operational station on the East Coast with a meteorological section and the metmen and meteorologists in the section provided the required services for the school.

SPECIALIST BCATP SCHOOLS

Central Flying School

In February 1940, the Central Flying School (CFS) and the Air Navigation School (ANS), both at Trenton, Ontario, were absorbed from the RCAF's Home War Establishment into the BCATP. There was a meteorological section at Trenton, serving both units, staffed by RCAF non-commissioned officers and airmen. Meteorologist Des Kennedy was posted to Trenton in May 1940 and he was replaced by Jim Leaver the following October. After the ANS moved to Rivers, towards the end of the year, there was neither a meteorologist nor a metman at the CFS for about a year on account of the general shortage of meteorological personnel. Metman Elmer Stevens was posted there from November 1941 until the summer of 1942 and later a meteorologist, Graham Potter, was posted there for eighteen months until February 1944. By mid-1945 four metmen were stationed at Trenton.

Flying Instructors Schools

In August 1942 three Flying Instructors Schools (FISs) were established. At Vulcan, Alberta, instructors were trained on single engine aircraft in use at some SFTSs. The Trenton, Ontario, school trained instructors on twin-engine aircraft that were used at other SFTSs while the instructors at Arnprior, Ontario, were trained on less powerful aircraft used at the EFTSs. The meteorological instruction at these schools consisted of thirty hours over a twelve-week period. This course was relatively advanced since the student instructors had already taken the SFTS meteorology course.

Weather briefings and lectures at ground school for the pilots on course at Trenton were given by the meteorologist or metmen at the Central Flying School. A metman was posted to the Vulcan school and another to Pearce, Alberta, in March 1943 when the school was moved there. A metman was posted to Arnprior for at least the first few months it was open in 1942.

Instrument Flying School

A metman was posted to the Instrument Flying School (IFS) which opened in April 1943 at Deseronto (Mohawk), Ontario. Weather forecasting and briefing were the main tasks of the metman at this school as no instruction in meteorology was regularly given. The school remained open after the termination of the BCATP when the RCAF absorbed it.

RADIO DIRECTION FINDING (RADAR) SCHOOLS

An RCAF radar school was established at Leaside, Ontario, and an RAF transferred radar school was established at Clinton, Ontario. The latter was taken over by the RCAF in July 1943. In the spring and summer of 1945 a metman (Don Perrie) was stationed at Clinton to assist in studying the use of radar in meteorology. The development of radar meteorology was in its infancy at that time although Perrie did telephone the Toronto Malton forecast office when the experimental radar detected severe thunderstorms over southern Ontario that summer.

OPERATIONAL TRAINING UNITS

Before the war, the Royal Air Force had organized special units for training pilots to fly operational aircraft before posting to front-line squadrons. In 1940 these were named Operational Training Units (OTUs). Four of these RAF units were transferred to Canada between June 1941 and June 1942. The first to arrive was located at Debert, Nova Scotia where operational training began in December 1941. Other RAF OTUs were located at Pennfield Ridge, New Brunswick; Greenwood, Nova Scotia; and at Patricia Bay, British Columbia, where crews were trained for Coastal Command squadrons.

The OTU syllabi called for eight weeks training for pilots and wireless operators/air gunners and eight for observers. In the final stages of their training, the aircrews, consisting of a pilot, an observer, and two wireless operator/air gunners flew together as a crew. The training at Debert prepared them for postings to Ferry Command units as trans-Atlantic ferry crews, which meant eight more weeks of training. Other crews were posted directly to the United Kingdom.

RCAF OTUs were opened as part of the BCATP in 1942 at Bagotville, Quebec, to give operational training to fighter pilots, and at Patricia Bay to train flying boat crews. By 1944 the RCAF staffed the RAF units as well as the Canadian units and opened a new OTU at

Boundary Bay, British Columbia, to train heavy bomber crews. Part of this unit moved to Abbotsford later that year and the Patricia Bay RAF unit moved to Comox. The OTUs were among the last units of the BCATP to disband. As is noted in chapters having to do with Eastern and Western Air Commands, OTU crews, both instructors and students, were called out on occasion to fly operational patrols.

All the OTUs had meteorological sections with meteorologists and metmen. The RAF and the RCAF provided airmen as assistants. Although the RAF/British Meteorological Office sent a few dozen metmen to staff some of their training stations in Canada this was not the case at the OTUs which were all staffed by Canadians. An excellent description of the Debert meteorological section in 1942 has been provided by a wartime metman, Graeme Cameron, who arrived there late in January that year.

> The Met Office, located one floor below the Operations room of the Control Tower, was manned by a meteorologist [metman] and an airman around the clock. This required a staff of four meteorologists, a corporal, and three airmen. The latter made hourly weather observations and reported them via the teletype machine to the Maritime network. They also plotted weather reports from the continental network on a weather map covering Canada and the United States every six hours. The "met" man drew up and analyzed these maps and prepared forecasts based on an analysis received from the Halifax Forecast Office, with occasional adjustments for local effects. They also gave refresher lectures to the aircrews, keeping them as practical as possible. . . .
>
> The meteorologist on duty would brief each aircrew on the weather conditions, cloud levels, wind directions and velocities expected along that day's flight path, and would give the navigator a card containing this information. The navigator in turn would enter on the card actual conditions encountered and would return this to us after the flight. To acquaint ourselves with the local weather conditions, we accompanied the aircrew occasionally on their training flights. These would be over the Gulf of St. Lawrence, the Bay of Fundy or south of Nova Scotia, often to Sable Island.[86]

The RAF flew Avro Anson and Lockheed Hudson aircraft in their training. Reid Dexter, a meteorologist, arrived to head the meteorological section in March 1942 and by 1944 he was assisted by a staff of four metmen and airmen assistants. In mid-1944 the RCAF took over the Debert OTU and the meteorological office remained open until early 1946.

RCAF FORECAST CENTRES

The Meteorological Division opened the Halifax aviation and naval forecast centre at the request of the Department of National Defence in September 1939. This office became part of Eastern Air Command and is dealt with in earlier chapters. Several months later the RCAF requested the Meteorological Division to open a similar forecast office for Western Air Command. Action was delayed because of the shortage of meteorologists but an office was opened in November 1940 in Victoria and when the WAC headquarters moved to Vancouver in 1943 the forecast office accompanied it. This forecast office has been described in earlier chapters dealing with Western Air Command. Both the EAC and WAC forecast offices existed primarily to serve RCAF operations and service was provided to the Royal Canadian Navy at both. A small forecast office was also set up at Rockcliffe airport in Ottawa to serve RCAF communications and transport flights; this office is described in Chapter 17.

When the BCATP schools began to open in 1940, mostly in southern Ontario and in the Prairie provinces, the meteorological sections were staffed by civilian meteorological officers or metmen who were "dependent forecasters." The task of providing guidance to the metmen was given to the District Aviation Forecast Offices (DAFOs), primarily those at Winnipeg and Toronto Malton and to a lesser extent at Lethbridge and Montreal. These DAFOs issued map analysis guidelines and RCAF regional forecasts. The DAFOs, established to provide meteorological services to Trans-Canada Airlines and other commercial aviation, became very busy as airway traffic increased markedly in the first two years of the war. By the end of 1942 it had become apparent that the DAFOs could no longer provide adequate services for the RCAF schools especially with their added needs on account of a marked increase in night flying training.

To better serve the southern Ontario schools and to relieve the Malton DAFO, the Meteorological Division considered three possibilities: expanding the DAFO, setting up a larger office at Ottawa Rockcliffe, and making an office at the Toronto Head Office where facilities already existed. Discussions indicated that there was not enough room to expand the office at Malton and an expanded Rockcliffe office would require a full staff of meteorologists, assistants, and teletypists. Despite the RCAF preference for the Rockcliffe office it was decided to put the office in Toronto where the senior meteorologists involved in training could monitor the forecasting activities.

The Toronto RCAF forecast office began operations on March 8, 1943, when it took over the issuing of all RCAF regional forecasts for southern Ontario. Two of these had been issued by the Montreal Dorval DAFO and the rest by Malton. Forecasts were issued four times a day from the office; meteorologists on staff were John Knox, Archie McCracken, and Gil Clark. Then, late in April, a circular signed by John Patterson was issued stating that, "Due to circumstances beyond the control of this office, it has been found impossible to maintain a sufficient staff of Grade 1 meteorologists at Toronto Head Office to continue the operation of the recently established RCAF forecast centre."[87] The RCAF schools were advised that the Malton DAFO would take over the RCAF forecasting duties on April 29 but no reasons were given for the change.

Although some consideration was given to setting up an RCAF forecast centre at Regina there was soon agreement that the new centre should be at Rivers, the home of the Central Navigation School. Ralph Anderson, a meteorologist, headed the meteorological section there and it was decided the new forecast office should be part of this section. The forecast centre began to operate on June 1, 1943, with meteorologists Einar Einarsson, Norm Powe, and Ray Walkden. Ralph Anderson took his turn in the forecast office as well as managing the combined office.

The Rivers forecast office was located in the ground instruction school but in July 1943 it was moved to more spacious quarters on the second floor of the control tower. The centre provided a twenty-four hour forecasting service for all RCAF stations in Manitoba and

Saskatchewan. Four synoptic maps were analyzed each day, eight radiosonde observations were plotted and upper air charts for five thousand and ten thousand feet (later for the 850 mb and 700 mb levels) were regularly plotted and analyzed.

Time did not permit the preparation of individual forecasts for each station so the region was divided into four districts for which separate forecasts were issued four times a day, each valid for eight hours. RCAF terms and definitions were used in wording the forecasts, which covered surface and upper winds, weather, visibility, clouds, notes, and a synopsis. The night shift forecaster was responsible for one to five briefings of aircrews between 0700 and 0830 hours at the end of his shift. The complement of forecasters at Rivers remained fairly constant over the months until it was disbanded in September 1945. In addition to the four meteorologists previously mentioned, Erling Anderson and Van Gordon arrived in mid-1944 and remained with the others until the office was disbanded.

WOMEN'S DIVISION

In 1941 the Royal Canadian Air Force organized a Women's Division (WD) and trained recruits to replace airmen in a number of trades. In the spring of 1942 it was decided that airwomen might well replace airmen as meteorological observers and the first WD course in this trade commenced at Toronto early in July 1942.

Many women selected for the first two meteorological observer courses had enlisted with the intention of asking for this trade as it appeared to them to be the most interesting trade available. The qualifications of those selected by the RCAF Trade Selection Board for training in the Meteorological Observer trade were high. All twenty-four women selected for the first course had at least junior matriculation standing, many were schoolteachers and a few had university degrees. Following a four-week period of basic training at Ottawa the women were sent to Toronto for "trade" training in meteorological work.

Each day for five weeks the women on course were marched from a Women's Division depot to the Royal Ontario Museum, near the Head Office of the Meteorological Division, where their course was conducted and supervised by a civilian meteorologist and assistants. Al Crocker, a meteorologist involved in training metmen and meteorologists, was initially responsible for the training. He was followed by Joan Griffiths, one of the few women to take the Intensive Course for metmen, and later by Murray Monsinger, a meteorologist on the headquarters staff.

Over the six weeks training period the WDs had thirty-five lecture hours in synoptic meteorology where the theories behind weather map analysis and forecasting were explained to them to enable them to better understand the practical work. They had ninety hours of instruction and practice in taking, coding, and transmitting weather observations including practice on teletype machines, and another ninety hours in learning to plot weather maps and diagrams. In this way they learned the methods and procedures which would be expected of them in RCAF meteorological sections. Examination today, nearly sixty years later, of the notes, practice map plotting exercises, and the final examination papers retained by Francess Halpenny, one of the members of Course #5, reveals both the intensity of the course and the quality of the WD trainees.

After graduation the WDs were posted, usually in groups of three or four, to training stations and operational bases where they replaced airmen. The first postings took place late in August 1942 to several Service Flying Training Schools since facilities for women were not yet available on operational bases. (Also, at first, women in the RCAF could not be posted to Newfoundland or anywhere else outside of Canada until their twenty-first birthday.) Usually an airman corporal was left at a station for a few months after the WDs arrived but within a short time all the meteorological assistants in the meteorological section were WDs with their own corporal. (In time a few WDs in the trade became sergeants.)

The introduction of the WDs into the activities of the BCATP meteorological sections was a marked success. The women usually brought more education and training to the job than the men had brought and most requested the trade rather than being ordered into it as most airmen had. The result was a more efficient meteorological office, which served the pilots and student aircrew better than the earlier offices staffed with airmen. However, after the high quality of the women at first, some difficulties arose in subsequent courses when a few recruits were unable to handle the training and others proved to be unsuitable for the work. On the other hand, since it was rarely possible to advance beyond the rank of corporal in the trade, some excellent WDs were lost to meteorology when they remustered to trades with better opportunities for promotion in the RCAF.

By February 1943 the need for meteorological observers decreased and it was the summer of 1944 before two more WD courses were given at Toronto. The last course, the tenth, was held in the fall of 1944 with twenty-six airwomen and one airman. Earlier, between November 1943 and November 1944, the Meteorological Division had given six observer courses for RCAF and RAF airmen in Toronto.

CLOSING DOWN

The November 1939 BCATP Agreement called for the plan to operate from April 29, 1940 until March 31, 1943. With the Plan working better, perhaps, than expected and with the need for aircrew still increasing, it was agreed at a conference in the spring of 1942 that the Plan should be extended to March 1945.

By mid-1943 the trained aircrew production peak was reached. In the month of January of that year there had been 1,574 aircrew graduates and by October the monthly totals had grown to 5,157 graduates and there was a large surplus of recruits. Historians have written that, by the fall of 1943, although the war was far from being over, the battle of aircrew training had been won.

Late in 1943, Canada and the United Kingdom, the principal partners in the plan, decided to disband a few Elementary and Service Flying Training Schools. By early 1944 the reduction was accelerated and it was planned to close thirty-two of the existing eighty-two schools. Despite the aircrew losses occurred in the Allied invasion of France in June 1944, the surplus of RAF and RCAF aircrew in Britain was such that Canada suspended aircrew recruiting late that month.

Canada and the United Kingdom agreed that Canadians would replace the RAF personnel at the transferred schools as soon as possible and that those schools would be the first to close. In November

1944 a decision was reached to close out the entire British Commonwealth Air Training Plan on March 31, 1945.

Just as the fast opening of the schools in 1940 and 1941 had caused John Patterson great difficulty in staffing them with metmen, the rapid closing of the schools in 1944 and 1945 also produced major problems for him. The Meteorological Division's recruiting and training program had been remarkably successful. From late 1940 until mid-1943 successive intensive training courses for metmen had been run one after another, sometimes with only a weekend between courses. The need for metmen had accelerated as the RCAF obtained increased meteorological section establishments but, by 1944, most schools had the metmen the training required. The need for metmen to replace RAF personnel at the transferred schools, the increasing needs at the operational stations on the coasts, and the loss of scores of metmen for training to become meteorologists (independent forecasters) made courses ten and eleven necessary in 1944 and 1945.

In January 1944 the Weyburn, Saskatchewan, school was the first RAF SFTS to close. However, an RCAF SFTS at Moncton was moved to that airport and there were no surplus metmen until it too closed in June of that year. By that time four more RAF schools in western Canada where Canadian metmen had been stationed were closed. In August, Kingston, the remaining RAF SFTS, where Royal Navy Fleet Air Arm pilots were trained, closed and the Aylmer RCAF SFTS moved there to continue the Royal Navy training. RAF meteorological personnel had staffed the Kingston meteorological section until Canadian metmen took over in October 1943. By December 1944, the surplus of metmen was such that five were posted to the Kingston school.

Besides the Aylmer school, four more RCAF SFTSs closed before the end of 1944 — Brantford, Macleod, Yorkton, and Dunnville — and their metmen were dispersed to other schools or, in some cases, allowed to resign if they were wanted in their prewar positions or if they had been accepted for study at a graduate school. Eleven SFTSs closed in March and April 1945 leaving only Kingston and Calgary to close in September and Camp Borden in March 1946. At several schools, such as Ottawa Uplands and Brandon, the RCAF requested that the meteorological section remain open with one or two metmen after the schools closed to serve itinerant aircraft.

Although a surplus of pilots occurred as early as the summer of 1944 there was still an overseas demand for other aircrew members and this kept the Air Observer Schools open longer than the SFTSs. For other reasons the Regina/Pearce and Prince Albert AOSs had closed prior to 1944 when the Edmonton school closed in August. London closed in December and the remainder in either March or April 1945. Since several of these schools were located at airports with civil meteorological offices there was no need to keep them open to provide weather forecasts and briefings for itinerant RCAF pilots.

Some meteorological offices remained open for weeks as the RCAF moved in other schools or units to replace those that had left. The Portage la Prairie AOS was succeeded by a third navigation school for a brief period, the Mount Hope ANS station became a Wireless School, the London AOS became a Mosquito aircraft delivery unit, and so on. The metman establishments at such schools were usually decreased but the offices remained open. Note has already been made of the posting in 1944 and 1945 of metmen to Elementary Flying Training Schools and Bombing and Gunnery Schools which absorbed metmen who might otherwise have become surplus. The need was real at these schools but metmen would not have been available had the need arisen in earlier years.

All metmen and meteorologists, in fact all employees hired by the Meteorological Division in the war years, were classed as temporary employees. In 1940, when the hiring of large numbers of metmen began, the Meteorological Division advised them that there was very little prospect of them obtaining permanent jobs in the postwar meteorological service. In July 1944, the meteorologists and meteorological assistants in war positions were advised by Patterson that he was making a survey of postwar requirements and asked them to advise him whether or not they wished to remain in meteorology. He told them he was still hard pressed to find sufficient staff to meet current RCAF requirements but he would try to arrange matters for the convenience of those who wished to leave that fall. A few metmen and meteorologists took advantage of the offer and returned to their prewar teaching positions or to graduate school.

As many SFTSs closed in the fall of 1944 the Meteorological Division moved metmen to the schools that remained open and to other

RCAF stations where a meteorological presence was still required in order to accommodate others who wished to leave the Division. The requirement for metmen at the AOSs did not decrease as rapidly since most of those schools remained open until the spring of 1945.

In June 1945 the RCAF called up twenty metmen who had volunteered for service in southeast Asia. In early August the other volunteers were advised that another group would soon be called up. But, within days, after the surrender of Japan, all planning and preparations for an RCAF Pacific Force stopped and the commissioned metmen soon returned to the Meteorological Division as civilians.

It was at this time that Andrew Thomson, assistant Controller of the Meteorological Division began to be concerned that the Division would soon have an excess of metmen. His circular of August 17 suggested that the metmen "would be well advised to obtain new positions." He wrote that the RCAF would still require a considerable staff of metmen so all who wanted release could not be accommodated at once but the national policy was that teachers would be released first if they so desired. Then, in October, Thomson wrote that there were still twenty-five more metmen on staff than could be employed in the post-war meteorological service.[88]

As the situation developed in the winter of 1945–46 only a very few metmen who wished to stay were not offered post-war positions. Three hundred and seventy-three metmen were trained in wartime; of this number, 108 were given advanced training to become meteorologists and 222 resigned at the end of the war. Forty-three metmen remained in the 1946 Meteorological Division and made meteorology their career. Two metmen were killed on duty during the war — Arthur Churchill Longman in an aircraft crash at Saskatoon in 1943 and Robert Bertram Munro in a Camp Borden traffic accident in 1944.

The resignation of meteorologists at the end of the war was not nearly so numerous as with the metmen. No more than thirty-five of those who took the advanced course during the war left the Division at the end. In summary, about 30% (112) of those trained as metmen remained in meteorology after the war as meteorological officers (metmen) and meteorologists.

[85] L.B. MacHattie, an unpublished manuscript on the history of Canadian meteorology in wartime, 1946.
[86] Personal archives, draft manuscript by Graeme Cameron, November 1986.
[87] AES, Archives, Circular 379 "Discontinuance of the YZO Analysis and Forecast Centre," April 27, 1943.
[88] AES, Archives, Circular 578, July 14, 1944; Circular 790, August 17, 1945; and Circular 823, October 11, 1945.

CHAPTER 15

FLYING TRAINING SCHOOLS

In 1945, during the last days of the BCATP, John Patterson requested that each training school meteorological section prepare and submit a report outlining the history of the section with an account of the ground school instruction and briefing activities over the period of operation. Unfortunately the personnel at only a few schools complied with this request. Of those that did submit a report, the metmen at the Brandon SFTS, the Malton AOS, and the Rivers Central Navigation School produced excellent reports and so those schools are featured in this chapter as representative schools of the wartime SFTSs, AOSs, and ANSs.

Where available, the names of the metmen posted to these three schools have been included as was done with the operational bases. While the identity of the metmen at the different schools in the final year of air training plan is known no lists are available for the first three years. Accordingly no attempt has been made to identify the metmen who were posted to the vast majority of the schools over the years. However, the names of all who took the various Intensive Training Courses in Meteorology are given in Appendix 3.

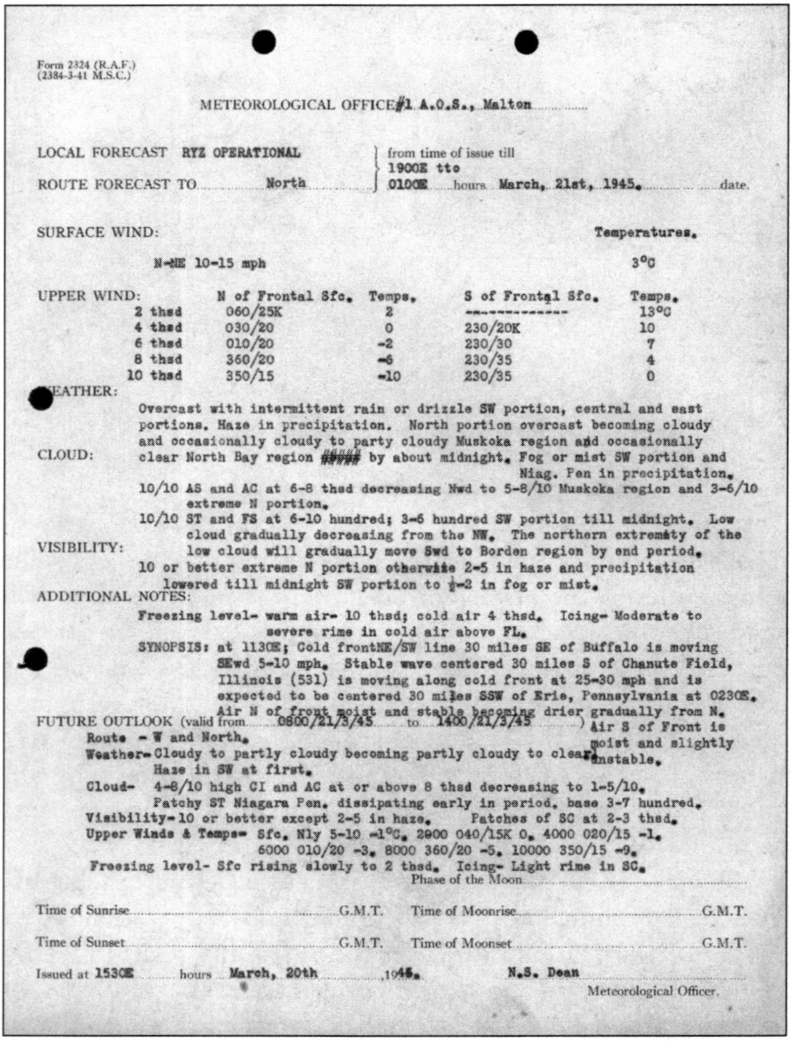

Form 2324 (R.A.F.)
(2384-3-41 M.S.C.)

METEOROLOGICAL OFFICE #1 A.O.S., Malton

LOCAL FORECAST RYZ OPERATIONAL
ROUTE FORECAST TO North
from time of issue till 1900E tto 0100E hours March, 21st, 1945. date.

SURFACE WIND: N-NE 10-15 mph

Temperatures. 3°C

UPPER WIND:

	N of Frontal Sfc.	Temps.	S of Frontal Sfc.	Temps.
2 thsd	060/25K	2	-------------	13°C
4 thsd	030/20	0	230/20K	10
6 thsd	010/20	-2	230/30	7
8 thsd	360/20	-6	230/35	4
10 thsd	350/15	-10	230/35	0

WEATHER: Overcast with intermittent rain or drizzle SW portion, central and east portions. Haze in precipitation. North portion overcast becoming cloudy and occasionally cloudy to party cloudy Muskoka region and occasionally clear North Bay region ##### by about midnight. Fog or mist SW portion and Niag. Pen in precipitation.

CLOUD: 10/10 AS and AC at 6-8 thsd decreasing Nwd to 5-8/10 Muskoka region and 3-6/10 extreme N portion.
10/10 ST and FS at 6-10 hundred; 3-6 hundred SW portion till midnight. Low cloud gradually decreasing from the NW. The northern extremity of the low cloud will gradually move Swd to Borden region by end period.

VISIBILITY: 10 or better extreme N portion otherwise 2-5 in haze and precipitation lowered till midnight SW portion to ½-2 in fog or mist.

ADDITIONAL NOTES:
Freezing level- warm air- 10 thsd; cold air 4 thsd. Icing- Moderate to severe rime in cold air above FL.
SYNOPSIS: at 1130E; Cold front NE/SW line 30 miles SE of Buffalo is moving SEwd 5-10 mph. Stable wave centered 30 miles S of Chanute Field, Illinois (531) is moving along cold front at 25-30 mph and is expected to be centered 30 miles SSW of Erie, Pennsylvania at 0230E. Air N of front moist and stable becoming drier gradually from N. Air S of Front is moist and slightly unstable.

FUTURE OUTLOOK (valid from 0800/21/3/45 to 1400/21/3/45)
Route - W and North.
Weather- Cloudy to partly cloudy becoming partly cloudy to clear. Haze in SW at first.
Cloud- 4-6/10 high CI and AC at or above 8 thsd decreasing to 1-5/10. Patchy ST Niagara Pen. dissipating early in period, base 3-7 hundred. Patches of SC at 2-3 thsd.
Visibility- 10 or better except 2-5 in haze.
Upper Winds & Temps- Sfc. Nly 5-10 -1°C. 2900 040/15K 0. 4000 020/15 -1. 6000 010/20 -3. 8000 360/20 -5. 10000 350/15 -9.
Freezing level- Sfc rising slowly to 2 thsd. Icing- Light rime in SC.

Phase of the Moon ....

Time of Sunrise .... G.M.T. Time of Moonrise .... G.M.T.

Time of Sunset .... G.M.T. Time of Moonset .... G.M.T.

Issued at 1530E hours March, 20th, 1945. N.S. Dean
Meteorological Officer.

FIGURE 8

A typical weather forecast as issued to student aircrew and instructors on Form 2342-3-41 M.S.C. at No. 1 AOS Malton, Ontario, on March 21, 1945.

FLYING TRAINING SCHOOLS

BRANDON SERVICE FLYING TRAINING SCHOOL

No. 12 Service Flying Training School, Royal Canadian Air Force, was located at the Brandon airport, eight kilometres north of the city of Brandon. The surrounding countryside is generally flat with Riding Mountain to the north. This school for advanced twin engine flying training used Cessna Crane and Avro Anson aircraft. Most flying was local with cross-country navigation training flights over southern Manitoba and southeast Saskatchewan within an area bounded by Weyburn, Saskatoon, Dauphin, Winnipeg, and the American border. The student pilots received their wings on graduation.

On May 16, 1941, metman Alex MacVicar opened the meteorological section at No. 12 SFTS. MacVicar, a 1937 graduate of the University of Western Ontario, had been hired by the Meteorological Division in November 1940 and graduated in February 1941 from the first Intensive Course in Meteorology for metmen. He had spent a month at the new Lethbridge District Aviation Forecast Office and six weeks in the meteorological section at the Macleod SFTS with observer-metman Harlan Thompson before coming to Brandon to open the new school. This was common practice when the first schools were being opened; within a year or so, when there already was a metman at each station, a new metman could be posted directly to a station off course without orientation at a civil office or another RCAF station.

Meteorological Office

The "Met Office" was located on the third floor of the control tower building, above the offices of the Commanding Officer and his staff and below the radio room. The flight controllers occupied the top glass-walled room. With a wooden map table, two kitchen tables, the teletype machine, and a few chairs the room had to be considered crowded. One side of the room was all windows and on the opposite side a counter separated the tower stairs from the office. The counter was convenient for visitors to the office but the stairs were a distraction to those working there.

Aircrew Trainees

Pilots were trained at Service Flying Training Schools. The students had taken a course for all prospective aircrew at an Initial Training

School and had been chosen for pilot training. Then they went to an Elementary Flying Training School to take instruction on small light aircraft like Tiger Moths and Stinsons. Here, after a number of hours of dual instruction they first flew "solo." Most were successful and graduated to an SFTS.

The first training course, number 28, arrived at Brandon on May 16, 1941, the same day as the metman arrived. (Courses were numbered consecutively in the BCATP from the first one in 1940; Course 28 students would have gone to several SFTSs.) In this instance the course intake at Brandon consisted of thirty-six Royal New Zealand Air Force students. Usually the courses were a mix of students from the RCAF, the RAF, and the Anzac air forces (Australian and New Zealand). As at other flying training schools half of the students flew in the morning and went to ground instruction school in the afternoon with the other half of the course doing it in reverse order.

Meteorological Section Staff

Two airmen assistants arrived on the same day as MacVicar, one with experience at another SFTS and one untrained who came directly from Manning Depot. The airmen assisted MacVicar in installing the observing equipment that had just arrived (except the barometer which was not found until September in the storeroom of the Officers' Mess). The teletype was installed at the end of May and some observations were commenced but it was late September before observations were taken routinely every day around the clock.

When MacVicar was posted to Toronto for the Advanced Course in October 1941 Rod McKenzie, fresh from the third short course, replaced him. As the metman manpower situation improved it became possible for the Meteorological Division to begin to staff each SFTS with two metmen and this took place at Brandon early in January 1942 when Jack Labelle arrived. Labelle had had a year's experience at the Winnipeg Air Observer School and so by the standards of the day was a veteran metman. In February, Bill Moore from the fourth short course arrived to replace McKenzie who was posted to Weyburn to help open the meteorological section in a transferred school, No. 41 SFTS. John Lauder replaced Moore in late 1942.

Two additional airmen were added to the staff in June 1942

allowing the office to be open from 0700 hours to 0200 hours each day when night flying was scheduled. The first airman NCO, a corporal, arrived in October. More than a year later, in October 1942, airwomen from the Women's Division (WDS) began to replace the airmen and for a few months the RCAF airmen/airwomen staff was larger than required. In December 1942 there were four airmen and three airwomen in the section and it was not until April 1943 that all the airmen had been posted out leaving a normal staff of four, all WDS.

In the fall of 1942 Mac Elsley arrived to provide holiday relief and when he left a third metman for the staff, L.B. Morrison, arrived fresh from a course in late December. From the arrival of Morrison to the closing of the school in March 1945, eight more metmen were to make their appearance and stay for varying periods of time at Brandon. Some like I.W. Findlay and Dick Bridgman came directly from a course and remained but a few weeks before moving on. Others, like A.J. Ducklow, A.H. Morrish, and Don Storr, came from other schools, were at Brandon for but a short time and left for other postings. Bill Mackie came from the Rivers Central Navigation School in late 1943 and remained until mid-1944. Finally D.D. Levi and M.A. Parker came in late 1944 and with John Lauder remained at the Brandon RCAF station for three or so months after the school closed to brief itinerant aircraft.

Ground School Instruction

When the Brandon school opened meteorology was still considered part of the navigation course and only eight hours were available for lectures. It would not be until August 1942 that the hours available for instruction in meteorology were increased to sixteen. The time allowed was increased again in 1944 to twenty-two hours and by the end of training the metmen at Brandon were given from twenty-five to thirty hours for meteorology instruction with each course.

The lecture program, with two morning and two afternoon lectures each week for each course, increased as new courses continued to arrive. By late 1941 it had become impossible for one metman to handle all the lecturing and still have sufficient time for forecasting and pre-flight briefing. In fact, this need was the major factor in getting the RCAF to increase the metman establishment to two throughout the system early in 1942. At full operation there were four courses at the Brandon school at any one time; usually the meteorology lectures were given in the first three months of each course.

As a rule the different metmen took alternate courses as they arrived. The chief ground school instructor was usually able to schedule the lectures in meteorology at times advantageous for the metman. At Brandon these times were between 1100 hours and noon and again between 1530 and 1630 hours. But this meant that there was no metman in the meteorological office during those periods. Having one metman do all the lecturing with the other two responsible for the forecasting and briefing rectified this situation.

At Brandon several of the metmen had been teachers before the war and this helped keep the teaching standards high. The Toronto Head Office provided examination papers but the students' papers were marked locally. The Brandon metmen recorded that the New Zealand and United Kingdom students, who had a good grounding in the subject at Initial Training School, usually showed the most interest in meteorology and made the best marks.

Besides the lectures, a few British Air Ministry films on meteorological subjects were shown, usually in the evening. Cloud pictures and diagrams illustrating hail formation, thunderstorms, and the development of icing at fronts were projected on a screen by an epidiascope (an early version of a still picture projector). Another aid used at all SFTSs was a glass air mass model sent from Toronto to use in lectures dealing with fronts.

The lecturer took a current map to each lecture and the students grouped around the front of the room for a brief map discussion before the lecture began. It was the conclusion of the Brandon metmen that "the more meteorology given to the students, the more interest and appreciation they showed in the subject."

Weather Maps and Forecasts

For the first several months, with two and then three airmen on staff, the office was open only during the daytime flying period from 0800 to 1800 hours. The airmen plotted two synoptic maps and these were analyzed by the metman. Upper air data were used only sporadically as only a few tephigrams were plotted during this period.

By May 1942, with the airman staff increased to five and two metmen on station, the office was kept open around the clock. Three synoptic charts, at 0730, 1330, and 1930 hours, and a local map at 0130 hours were plotted and analyzed each day. Eight tephigrams were plotted twice a day, a 10,000-foot (3,048 metres) map was plotted, and cross-sections prepared for the briefing sessions. After March 1943 a metman was on duty from 0700 to 0200 hours the next morning and would remain on duty as long as night flying continued in doubtful weather.

The metmen at Brandon had the same difficulty as those at the other SFTSs in Saskatchewan and Manitoba. Being dependent forecasters their instructions were to rely on the basic synoptic analysis and regional forecasts from the Winnipeg DAFO before completing their own analysis and issuing forecasts. But the RCAF flying instructors required forecasts before the Winnipeg analysis usually became available. Fortunately the Lethbridge DAFO analysis, also carried on the teletype circuit, was received in time to assist the Brandon metmen in their analysis and local forecasting. After June 1943, the Brandon metmen found forecasts from the Rivers Training Command forecast office to be better suited to their needs than the Winnipeg forecasts had been.

In 1941, when the only metman in the section was lecturing at ground school, the airman on duty would simply copy the Winnipeg regional RCAF forecast onto forms for distribution to the flights. When the office was fully staffed forecasts were delivered to the various flight offices at 0700 and 1200 hours and when night flying was scheduled, new forecasts were delivered at 1800 hours and if still flying, again at 2359 hours. Forecasts were also delivered to the Chief Flying Instructor, the ground school, the control tower, and other interested offices on the station.

The forecasts were typed onto a standard form (Form 2384 MSC) showing the forecast surface wind, upper winds at 5, 10, and 20 thousand feet, the weather, cloud, visibility and freezing level, additional notes and a further outlook. When conditions that might be encountered on cross-country navigation flights were different from those forecast for the station, this information was given in the additional notes. A further outlook section completed the standard forecast.

The forecasts were usually hand-written by the metman on duty and typed by an airman. A gelatinous Ditto machine was used to duplicate copies.

Along with the forecasts, weather reports from nearby stations were provided to the flights and offices. This information was given in tabular form at first and later in map form. Wallboards or cases were available in which to place the maps and forecasts in the different flight offices. Forecasts for the First Canadian Parachute Battalion at Camp Shilo were made twice daily at 0600 and 1800 and sent by telephone. Verbal forecasts were provided for itinerant aircraft but written forecasts were given if the flight was of long duration and/or distance.

Weather Briefings

There was no organized schedule of briefings for the first year or so at the station. Pilots and students came individually to the office for information. Mass briefings of instructors and students were begun for the Navigation Flight in July 1943 and in December the briefing program was extended with briefings in each hangar at 0800 hours. Mass briefings for night flying began in February 1944. Beginning in September 1944, daily mass briefings for all pilots and students scheduled to fly that day were held in the Ground Instruction School and, in a few weeks, the briefings were moved to the drill hall.

Using weather maps, the metman gave a brief review of the current synoptic situation with emphasis on the surrounding area, reported on the current weather at nearby stations, and outlined the forecasts for both the local area and for the scheduled cross-country routes. The briefing required two large blackboard maps at first but later an epidiascope was used to project the maps on a screen. Two metmen were required for these morning briefings, one coming on duty at 0630 hours and the other at 0730 hours to prepare for the 0800 briefings. There also were briefing sessions each day in the ground school for students scheduled to fly that night and later briefings in the meteorological office for students flying on late night shifts.

Observing Program

The first weather observations were not recorded until July 1941. In September, with more airmen on staff, a twenty-four hour observing program was begun. Because of the number of reporting stations on the circuit Brandon only transmitted the odd hour reports. Four synoptic observations were taken every day as well as "specials" when the weather deteriorated.

The airmen worked in three shifts, 0001–0745 hours, 0745–1700 hours, and 1700–2359 hours. As well as taking observations on each shift, an airman's work included such office duties as answering the telephone, plotting maps and tephigrams, making tables or maps showing the current weather at nearby airfields, tracing maps for the Navigation Flight and Ground Instruction School, and distributing the forecasts, reports and maps to the flights. Another duty of the observers was sending flight plans and flight arrival messages. (Before departing for another airfield a pilot was responsible for filing a "flight plan" giving aircraft type and number, the destination, and the estimated time of arrival.) This information was provided to the meteorological office by the officer in the control tower who was, in turn, advised when the aircraft had arrived at the other station as well as notice of any incoming flights.

On October 18, 1942, the first Women's Division assistants arrived and for some time the staff included both airmen and airwomen until the last airman was posted out in April 1943. Although the establishment called for four WDs on occasion only three airwomen were available to staff the office during most of the remainder of the year. In January 1944 the staff was increased to five airwomen.

MALTON AIR OBSERVERS SCHOOL

No. 1 Air Observers School (AOS) was located at the Toronto Malton airport (now known as Pearson airport). The airport is about 19 kilometres west-northwest of downtown Toronto and about 16 kilometres north of Lake Ontario. The surrounding countryside is generally flat with a gentle rise northwards from the lake to the Niagara Escarpment 24 kilometres from the airport.

The Malton AOS opened in May 1940, the first AOS in the British Commonwealth Air Training Plan, and was operated by Dominion Skyways (Training) Limited. Most aircrew training was with Avro Anson aircraft flown by civilian pilots; the training flights ranged over most of southern Ontario avoiding, of course, the United States. For the first eighteen months the AOS shared the airport with an Elementary Flying Training School. Trans-Canada Airlines used the airport for its regularly scheduled commercial flights and Victory Aircraft Ltd., manufacturers of the Lancaster heavy bombers and other warplanes, made constant use of the airfield.

Meteorological Office

The "Met office" was located in the Ground Instruction School and was used by both the metmen instructors and forecasters until March 1944. That month the forecast part of the office was moved to #3 hangar.

Unlike the meteorological offices in control towers at the Service Flying Training Schools this AOS office did not have a sweeping view of the airfield. With the metmen instructors, the forecasters and the assistants all using the office, it was somewhat crowded. Space along the walls was given over to a plotting table, the teletype, a forecast desk, a table for the Ditto machine, a typewriter, a phone, and a window. Near the entrance to the room was a large counter for displaying the current maps and forecasts used in consultations.

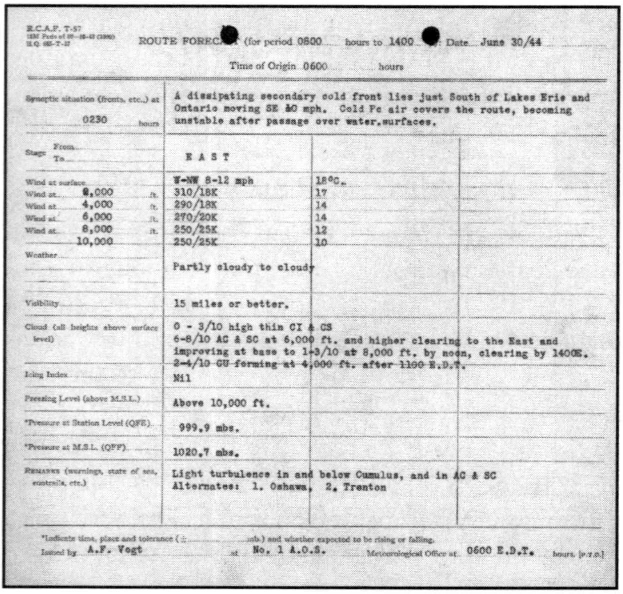

R.C.A.F. T-57

ROUTE FORECA[illegible] (for period 0800 hours to 1400 [illegible]: Date June 30/44

Time of Origin 0600 hours

Synoptic situation (fronts, etc.) at 0230 hours	A dissipating secondary cold front lies just South of Lakes Erie and Ontario moving SE 20 mph. Cold Pc air covers the route, becoming unstable after passage over water.surfaces.		
Stage From / To	EAST		
Wind at surface	W-NW 8-12 mph	18°C.	
Wind at 2,000 ft.	310/18K	17	
Wind at 4,000 ft.	290/18K	14	
Wind at 6,000 ft.	270/20K	14	
Wind at 8,000 ft.	250/25K	12	
10,000	250/25K	10	
Weather	Partly cloudy to cloudy		
Visibility	15 miles or better.		
Cloud (all heights above surface level)	0 - 3/10 high thin CI & CS 6-8/10 AC & SC at 6,000 ft. and higher clearing to the East and improving at base to 1-3/10 at 8,000 ft. by noon, clearing by 1400E. ~~2-4/10 CU forming at 4,000 ft. after 1100 E.D.T.~~		
Icing Index	Nil		
Freezing Level (above M.S.L.)	Above 10,000 ft.		
*Pressure at Station Level (QFE)	999.9 mbs.		
*Pressure at M.S.L. (QFF)	1020.7 mbs.		
Remarks (warnings, state of sea, contrails, etc.)	Light turbulence in and below Cumulus, and in AC & SC Alternates: 1. Oshawa, 2. Trenton		

*Indicate time, place and tolerance (± mb.) and whether expected to be rising or falling.

Issued by A.F. Vogt at No. 1 A.O.S. Meteorological Office at 0600 E.D.T. hours. [P.T.O.]

FIGURE 9

A typical route forecast issued to student aircrew and instructors on Form RCAF T-57 at No. 1 AOS Malton, Ontario, on June 30, 1944.

Aircrew Trainees

At first only air observers were trained at Malton. In mid-1942, new aircrew categories were introduced and after that summer air observers, air navigators, and air bombers were trained. The various courses consisted of representatives of the RCAF, RAF, RAF (Polish), Royal Australian Air Force, the Royal New Zealand Air Force, and the Free French Air Force. American student airmen made up a minor segment of the RCAF trainees before the United States entered the war.

The students were flown, usually in Anson aircraft, by civilian pilots who took flight route instruction from the navigation students and on bombing exercises from the student bombers. The crew of each aircraft would usually consist of the pilot, two student air navigators, one student air bomber, and one student wireless air gunner. Normally there were twenty-six aircraft in a flight carrying out an exercise.

Meteorological Section Staff

The Malton school was unique amongst the BCATP schools in that no civilian meteorological staff was posted there for nearly a year. With the shortage of metmen and the co-existence of a District Aviation Forecast Office to serve Trans-Canada Airlines and other commercial aviation companies, the DAFO resources were utilized to the full at Malton. Fred Turnbull, the DAFO officer-in-charge, took responsibility for all forecasting, briefing, and lecturing duties required at the school and in this work he was assisted by observer G.K. Boyd and, for the first two months, by Fred Patterson.

Fred Turnbull was relieved of his AOS duties when a metman, Edward Longhurst, arrived in April 1941 from the first intensive course. Longhurst was the sole metman at No. 1 AOS until February 1942 when a new metman, Bill Jarmain, arrived. In a few weeks W.G. Wellington replaced him and John Miller replaced Wellington later that spring. The establishment of metmen was raised to three in September 1942 when D.A. Moddle arrived. Miller was replaced by N.H. McLelland in December and with the arrival of Charlie Goodbrand in June 1943 there were four metmen on staff. That level remained until the end of the year although several staff changes took place with the arrival of J.W. Solman, H.D. Skirrow, and I. Sheichet replacing those who left for other posts.

Longhurst departed in December 1944 and Moddle became the officer-in-charge. With the increased supply of metmen available the establishment at Malton was raised to five metmen in March 1944, to six in April, and seven in May 1944 although from that month until the end of training in March 1945 there were eight and later seven metmen in the meteorological section. After Moddle was posted out in August 1944 he was succeeded as officer-in-charge by W.R. Cunnington who was followed by Percy Saltzman in January 1945. Other metmen, not listed earlier, who served at Malton in 1944 and 1945 were Art Vogt, F.J. Turner, Ted Hamilton, Eric Dexter, J.W. Forrest, N.S. Dean, Thomas Hull, David Shales, Tom Nixon, T.H. Lopson and O.K. VanSickle. There were in all twenty-four metmen who were stationed at this AOS during the war.

The observers at the DAFO were completely responsible for the Malton observing program and so not as many airmen were required at this school as at most other AOSs. When a separate AOS meteorological section was set-up in March 1941, three RCAF airmen were assigned to assist Longhurst. By the time the training program was in full operation in early 1943 and the meteorological section was providing more forecasting and briefing services than had been the case in the early months, there were six airmen assistants including a sergeant and a corporal.

Ground School Instruction

From the opening of the school until July 1942 forty-eight observer courses were trained. Besides meteorology the ground school instruction included navigation, aircraft recognition, armament, signals, photography, and airmanship. The training schedules and syllabi were changed from time to time. For most courses prior to the summer of 1942, the observer course lasted fourteen weeks during which thirty-five to forty hours were devoted to meteorology. Some of these hours were spent viewing aviation meteorology films and some practical weather map work. The students graduated with an "O" wing. By August 1940 there were three observer courses in training that meant fifteen hours a week lecturing for the metman and this left little time for forecasting and briefing activities.

With the training change in 1942 when the air observer classification was dropped, the air navigators were trained in a sixteen-week course and the air bombers, a six-week course. The navigator course was soon extended to twenty weeks and at the end of 1944 the bomber course was extended to ten weeks. In full operation the student navigators received thirty-five hours of meteorology and the bombers twenty hours.

As at all BCATP schools each course at Malton was divided with one half flying in the morning and the other half in the afternoon. When not flying, the students went to ground school and this meant that the metmen had to give every lecture twice. By 1943, when in full operation, there were ten double courses of student air navigators and five double courses of air bombers at the school at the same time. With fifty-two students in each course this meant the number of student aircrew at any one time at Malton totalled 520 navigation trainees and 260 bomber trainees.

The navigators' meteorology course was very substantial and dealt with the composition of the atmosphere, clouds, air masses and fronts, the heat balance of the atmosphere, wind, precipitation, and those elements hazardous to aviation — ice accretion, thunderstorms, and fog. The bombers' course was shorter and did not include map analysis and not as many films were shown.

Each lecture hour was fifty minutes in length. Lectures were given in the 0900 to 1700 hour period daily except Sundays. Three of the eight metmen on strength were detailed to instruction duties. Two metmen handled each double course of navigators and in a two-week period each would have twenty-eight hours of lectures, two sets of tests to mark, two periods of films, and one set of final examinations to set and mark. One metman handled each bomber course and in each two-week period had thirty-six lectures, two sets of tests, two periods of films, and two sets of final examinations.

As well as the lecture programs for the students the metmen, on occasion, gave refresher courses for navigation instructors, senior pilots, and new staff pilots. When wireless airgunners were at the school a short course of six meteorology lectures was given to them.

Weather Maps and Forecasts

As earlier noted, the Malton District Aviation Forecast Office provided all required forecasts for the school in the early months. These forecasts were tailored to the AOS requirements by the school metman. Forty-five minutes prior to briefing time the Duty Pilot, the Duty Flight Lieutenant, and the AOS duty metman went to the DAFO where they were briefed orally. Returning to his office the metman used this information, his map, and the regular regional RCAF forecast (prepared by the DAFO) to compose and type out a route forecast for the training flights scheduled for that day. The AOS airman made copies for all the aircraft on the flying detail and distributed them along with map tracings for briefing and display purposes at the school. Then, the metman kept in touch by telephone with the DAFO to check for any changes in the observed weather or forecasts until the aircraft returned.

This cumbersome system continued until March 1944 when an AOS forecast office was organized and located in #3 hangar. (The AOS instruction office remained in the ground school building.) Beginning that month an AOS metman was on duty twenty-four hours a day (shifts were 2359–0900 hours, 0900–1700, and 1700–2359 hours). All analyses and forecasts were prepared using the DAFO analysis and regional forecasts.

The night shift was a busy one. The 0230 hour synoptic map, two upper level maps, and several radiosonde ascents were plotted by the airman. The metman analyzed the maps and produced a "Weather for Today" general forecast, an operational forecast for the afternoon which covered surface and upper level winds, weather, clouds, visibility, and other items of concern to fliers. Special forecasts for the day's bombing exercises were also prepared. These were all issued at 0700 hours.

Two synoptic maps were plotted and analyzed on the day shift and three operational forecasts were issued for the flying exercises that evening, that night, and next morning. The evening shift was not as busy; one synoptic map was analyzed but no forecasts were issued. There were large notice boards positioned around the school for displaying the current and preceding weather maps and several boards for posting the current forecast. Forecasting services were also provided for the de Haviland Aircraft Co. test flights at nearby Downsview airport and for various station offices and special flights having to do with photography and bombing exercises.

Weather Briefings

By the time the school was operating at full capacity in 1943 an extensive consultation and briefing schedule had been developed. About an hour and a half before the first aircraft in the day's exercise took off the metman met with the Flight Commander and the Duty Flight Lieutenants to help them decide on the routes and heights to be flown that day. It was at this conference that a decision was made to "wash out" flying for the day if the forecast was unfavourable.

The "Stage One Briefings" were held in a classroom, usually two or three hours before flight time and lasted for nearly an hour. The navigation instructor in charge of the exercise outlined the details of the exercise and had the students make preparations for the flight. The metman was given fifteen minutes at each such briefing; he used a blackboard, maps, and cross-sections to explain the synoptic situation to the students who were invited the ask questions.

The Stage Two Briefings were held thirty-five or so minutes before takeoff and were factual in content with discussion. The metman had about five minutes to outline the forecast while displaying maps and cross-sections. Much attention was given to the T-57 forms on which the forecast winds, weather, visibility, icing possibilities and so on were entered. On the reverse side of the form the students were instructed to record the actual weather they encountered.

Observing Program

As previously noted the Malton DAFO observing staff was responsible for the airport weather observations. Some limited observations were taken by the AOS airmen at times for specific purposes but were limited to using the barometer, barograph, and sling psychrometer.

RIVERS NAVIGATION SCHOOL

The RCAF Air Navigation School at Trenton was absorbed into the BCATP in April 1940 and in November of that year it was moved to Rivers, Manitoba. The meteorological requirements of the Trenton ANS had been the main reason for Des Kennedy's posting to that station in May 1940. When Kennedy was brought back to Toronto to assist in the metman-training program, Jim Leaver replaced him. In turn, Leaver was also recalled to Toronto but before he went he travelled to Rivers with the school. Another meteorologist, Ralph Anderson, accompanied him and took over the Rivers meteorological section in a month or so.

Before the war Rivers was a new secondary airport on the transcontinental airway between Winnipeg and Regina where the radio range operators took the weather observations. Because of this no observing equipment was installed but within a few months, the airmen at the school commenced a pilot balloon observing program.

Meteorological Office

A small meteorological office with a teletype, a map plotting table, an analysis table, and a desk was set up in a small flight room in a hangar. A room in a barrack block was used as a classroom until the new Ground Instruction School building was completed early in 1941 and all classes were moved to that building. At the same time the meteorological office was moved to a large classroom in the new building.

Aircrew Trainees

Training began in December with four specialist navigation courses requiring instruction in meteorology. One navigation course was for SFTS graduates who were to become navigation instructors; their course called for twenty hours of meteorology in six weeks. Two courses were given for university graduates who would become ground school navigation specialists at various schools in the BCATP; this course called for thirty hours over an eight-week period. The fourth course was an extensive one for pilots who were expected to later hold responsible senior positions in the RCAF. Both theoretical and practical aspects of meteorology were stressed in forty-three hours of meteorology given in each twelve-week course.

In 1941 the school continued to expand and the flying program increased, both day and especially night flying, since all instruction in astro navigation was given at this school. Due to the congestion, part of the training was moved in July 1941 to a new school, No. 2 ANS at Pennfield Ridge, New Brunswick. Ade Lenahan, then a metman, was posted to Rivers for six weeks before the move and became the officer-in-charge of the meteorological section at the new school.

The year 1942 became an eventful one for the Rivers ANS. In May the Pennfield Ridge school moved back to Rivers and the combined school was renamed the Central Navigation School. A few months later, in September, the Prince Albert AOS was closed and its training program taken over by Rivers. At about the same time the RCAF decided that all astro navigation training should be done at the AOSs. This decreased the amount of night flying at Rivers and thus the number of nights it was necessary to have a metman on duty.

Ground School Instruction

In June 1944, when the Rivers school was in full operation, the students were divided into three squadrons — two squadrons consisting of ten navigator classes each and a third consisting of ten air bomber classes and various specialist navigation classes. Eighty hours a week of lecturing in ground school were scheduled and this required four metmen who devoted themselves entirely to this work. One metman instructor was assigned to lecture to each of the two navigation squadrons; each group had twenty hours of meteorology a week for eighteen weeks. Two metman instructors were assigned to the third squadron where about thirty hours of instruction were required for the six-week air bomber courses and a few hours for each specialist course which lasted either six or twelve weeks. Besides having about four lectures a day each metman marked the T-57 reports from the students, marked the examinations, and did the Stage One briefings. The latter required about four hours a day to prepare and deliver the briefing.

Weather Forecasts and Briefings

From the opening the Rivers school had a sufficient number of RCAF airmen to plot four synoptic weather maps and the necessary tephigrams each day. Anderson was responsible for analyzing the weather maps and he independently prepared local and route forecasts for the training exercises as well as lecturing in ground school. Prior to June 1943 the Winnipeg DAFO had provided an RCAF regional forecast for the schools in Manitoba and Saskatchewan and both Winnipeg and Lethbridge issued airways forecasts which were useful to Anderson in preparing his forecasts. Anderson did all the forecasting until other meteorologists joined the staff in early 1943 and a new RCAF forecast centre was organized at Rivers (this centre has been described in another chapter).

Forecasts were distributed to the ground school offices and the flights during the first year or so. Routine navigation flights made up the bulk of the flying and the navigation instructor in charge of a class needed some information regarding the possibility of flying twelve to twenty-four hours in advance. He usually visited the meteorological office to get these forecasts. There was no real briefing program as such at first and the officer in charge of flying merely called the meteorological office before each flight to "check the weather." By January 1941, when ten to fifteen aircraft were involved, the pilots went to the office for a verbal briefing before each flight. It was April 1941 before the first real briefings took place. These were held in a special "Briefing Room" in ground school where the metman used the latest weather map to brief pilots and students.

By the fall of 1943, after a separate forecast office had begun to operate, more elaborate briefings were commenced. A metman gave the Stage One briefings in ground school at 1130 hours for flying in the afternoon and another at 1630 for those flying at night and the next morning. Preceding this briefing there had been a pre-flight conference involving the duty meteorologist, the metman briefing officer, and the RCAF officers in charge of flying and of the particular navigation course that was to fly. Here the decision to fly or not was made, and if flying, the routes and heights of the flight were determined. By the summer of 1944 as many as sixty-three Ansons would participate in a particular night exercise.

A meteorologist from the forecast centre gave the Stage Two briefing. Two large blackboards were used, one for a map and the other for a vertical cross-section of the atmosphere. The area covered in the horizontal map extended from the Great Lakes westward to the Rocky Mountains while the cross-section showed conditions at even levels up to 10,000 feet. (By 1945 the blackboards were abandoned for delineascope projections of maps and cross-sections.)

In all the briefings much attention was given to the T-57 forms. Before a flight, one side of each form was completed by the students from the forecast given during the briefing but by 1943 it was found better to issue a form to each student with the forecast reproduced on it. Then, during the flight it was the responsibility of each student to take actual observations on the reverse side. Completion of these forms was a factor in assigning marks to the students in the course.

Meteorological Section Staff

From the beginning, flying exercises were scheduled for both daytime and nighttime and it quickly became evident that one metman could not handle all the lecturing as well as the forecasting and briefing. In late January 1941, Don McGeary, an experienced Regina observer, was posted to Rivers to assist Anderson and when he left in March to open a meteorological office at the new Dauphin SFTS he was replaced by William Buscombe, a graduate of the first metman course. In October Karl Buckthought replaced Buscombe when the latter was recalled to Toronto to take the advanced course. Then, after the training programs at Pennfield Ridge and Portage la Prairie were moved to Rivers there was a marked increase in the amount of instruction in meteorology required and three metmen, Ted Munn, Gordon Allison, and A.J. Ducklow, were posted to the school.

In mid-1943 the metmen doing the instruction were Harvey Johnston, Art Lamont and Bill Mackie, and by the spring of 1944 the establishment was increased to four. Johnston remained at the school until it closed late in the summer of 1945. Other metmen at Rivers for appreciable periods during these years were G. Donaldson, D.E. Rivers, J.G. Moe, Fred Burbidge, Eric Dexter, Gordon McKay, and Des Wright. Metmen there for less than three months each in 1944 and 1945 included J.H. Crowley, Bill Jarmain, G.M. Busche, N. Hunchak, Fred Ingall, and B.V. Benedictson.

Five RCAF airmen arrived with the meteorologists and in time the total establishment of RCAF assistants was raised to one sergeant, one WD corporal, and four WD airwomen who began to arrive on station in January 1943.

CHAPTER 16

THE ARMED FORCES

By a large margin, most of the meteorological services provided to the Armed Forces in wartime were for the Royal Canadian Air Force. The Royal Canadian Navy was also a significant user of meteorological observations and forecasts but relations with the navy were never as easy and satisfactory to the Meteorological Division, as were relations with the air force. Except for providing personnel for secret research at Suffield and cold weather exercises, about the only other need of the Canadian Army was the artillery's need for ballistic winds.

Besides describing the services provided for the navy and army this chapter includes the Meteorological Division's other wartime activities with the Armed Forces — weather reconnaissance flights, Japanese fire balloons, weatherships, and the manufacture of special, secret instruments.

ROYAL CANADIAN NAVY

The Meteorological Division had little contact with the Royal Canadian Navy before World War II. In 1934, the navy had asked that weather forecasts and synoptic reports from a selected list of twenty-one stations in eastern Canada and the United States be sent daily to the Halifax Naval Radio Station for broadcast to the ships at sea.

The navy was primarily interested in the northwest Atlantic — the area westward from longitude 35 degrees West between latitudes 40 and 65 degrees North — and wanted this information from June through October each year.

With the outbreak of war, the naval staff used a naval code for transmitting to the ships and these transmissions became known as the Fleet Synoptic Messages. Large ships carrying meteorological officers required the complete messages while the smaller ships wanted only the forecasts. The large ships could also pick up weather broadcasts from the United States Navy's Arlington, Virginia, wireless station. By the end of November 1939, the Toronto Head Office was also providing forecasts twice daily to naval control officers at Saint John, New Brunswick, and Sydney, Nova Scotia.

In early 1940, in order to transmit more information over longer distances to the Royal Navy aircraft carriers and other large ships, naval authorities began to use the more powerful Louisbourg wireless station instead of the one at Halifax. Synoptic observations were coded and packaged at Toronto for transmission with forecasts prepared at the new Halifax forecast office. At about that time, the British Admiralty requested that synoptic messages for stations in eastern Canada be cabled daily to the United Kingdom for broadcast to the Royal Navy's Home Fleet. In 1942, arrangements were made for the Arlington station to also broadcast the Canadian Fleet Synoptic Messages and in 1943 a much more detailed analysis code, the Inter-Allied Fleet Analysis Code (IAFAC), was brought into use.

In December 1939, on the West Coast, meteorological and navy officials at Esquimalt and Victoria, British Columbia, arranged for coded messages to be sent to Royal Canadian Navy ships operating in the Pacific Ocean should this become necessary. When plain language broadcast of forecasts and bulletins ceased after Pearl Harbor this coding was done. In August 1942 the Esquimalt Naval Radio Station began broadcasting coded analyses and forecasts but not synoptic reports. This was probably because the United States Naval Station at San Francisco was broadcasting synoptic and upper air data for Pacific coast stations four times daily. In 1943, the new IAFAC code came into use on the Pacific as well as the Atlantic coasts.

THE ARMED FORCES

Naval Officers

On shore, particularly on the Atlantic coast, cooperation between the Royal Canadian Navy (RCN) and the Meteorological Division suffered from the determination of senior naval officers that their weather briefings must be provided by a naval officer and not by a civilian. This was in complete contrast to the position taken by the Royal Canadian Air Force where senior officers had no problem with direct contact with civilian meteorologists and metmen. The RCN's attitude was in line with the conventional position taken by the Royal Navy although that organization had its own Naval Meteorological Service with trained Naval Meteorological Officers. (Even so, in some parts of the world, the RN was forced to rely on Royal Air Force meteorological offices.) This Canadian navy position, one also taken by the United States Army Air Force, was to plague relations with the Meteorological Division throughout the war.

In the early war months the naval officers at the Halifax Dockyards received maps, forecasts, and weather reports daily from the Halifax forecast office by courier and telephone. Then, in December 1941, the RCN established meteorological officer positions at Ottawa, Halifax, Esquimalt and St. John's, Newfoundland, and early in 1942 began recruiting to fill the positions from the ranks of the Meteorological Division's metmen. John Patterson did not wish to lose trained metmen nor to fragment his Division so countered with the suggestion that the navy should recruit from officers returned to shore duty after experience at sea. Patterson volunteered that the Meteorological Division would give the chosen officers training in meteorology at the Halifax forecast centre. But the RCN claimed they could not spare any officers for this work and pressed the Meteorological Division to release the metmen they wanted. Patterson refused and again urged the navy to accept meteorological services directly from his Division.

But, the navy continued to insist that a naval officer, with naval traditions, was necessary to interpret weather information to the senior officers and that the navy could not have civilians in the naval operations room. Patterson wondered how a man could acquire naval traditions by taking a short course and considered the Navy's proposal that a meteorological officer be briefed in a forecast office and then relay the briefing to captains and other senior officers in their offices, to be most unsatisfactory. The RCN took the matter to the Wartime Bureau of Technical Personnel who ruled that although the armed forces could not recruit any metmen or meteorologists until they were cleared by the Meteorological Division, they were free to take men who had resigned from the Division. Patterson was ordered to accept the Bureau's decision and the resignation of metmen who intended to enlist in the Royal Canadian Navy.

Atlantic Coast

The first metman to resign from the Meteorological Division and take a commission in the RCN was Brock McElheran, a metman graduate of the first intensive training course in 1940–41. McElheran had been posted to the Rockcliffe RCAF station at Ottawa in the fall of 1941 where he joined the naval reserve and began officer training in the evenings. In May 1942 he was called up for active service, resigned from the Meteorological Service, and posted to Nova Scotia for more naval training. He was soon sent back to Ottawa and was employed in non-meteorological work for a year or so. Then, in June 1943, he was again posted to the Halifax naval Operations Room as Staff Officer (Meteorology). McElheran has described his duties as follows:

> My new job was to act as liaison between the navy and the civilian meteorological office. . . . Each day I was given a tracing of the weathermap and was briefed by the duty forecaster. I then took the map around the operations room staff from the admiral on down, after which I toured the dockyard. I visited the escort vessels as they came in, corrected their aneroid barometers . . . and had a chat with the navigation officer. I would ask for comments on our forecasts and encourage him to add coded weather reports to any messages sent from sea as we had no other means of knowing the weather conditions beyond our own visibility limits.[89]

In Ottawa, during the last half of 1942, more difficulties arose between the Meteorological Division and the RCN. J.H. Meek, another metman, who had graduated from the second short course in June 1941, was sent to the RCAF's Ottawa Rockcliffe station to join McElheran. The Rockcliffe meteorological section served communications and transport flying at the RCAF headquarters. In 1942, with McElheran in the navy and gone to Halifax, the Ottawa Naval Headquarters asked Meek to provide them with the same information the RCAF was receiving — a tracing of the morning's weather map and a number of morning weather reports from Canadian and American stations. John Patterson ordered the practice to cease since he felt the navy might be attempting to make forecasts for the Atlantic from this information which was totally inadequate for that purpose.

Meek stopped, the navy complained and after a request for the service was made at the Deputy Minister level, Patterson was forced to let Meek give the naval officers what they wanted. By November 1942, the Rockcliffe office was providing two map tracings a day, copies of hourly weather reports, and copies of the "Fleet Forecasts for Operations" prepared for Halifax and Esquimalt. The navy officers also had daily telephone discussions about the East Coast weather situation with the metman.

By 1943, when a forecast centre began to operate at Rockcliffe, the meteorologists provided additional services to the Ottawa Naval Headquarters. At Halifax, that spring, daily conferences were begun in the Naval Operations Section each morning, which were attended by senior RCN and RCAF officers. At each conference the Halifax duty meteorologist described actual and forecast weather conditions over areas where aerial coverage of convoys was planned. The forecasts covered periods up to forty hours in the future; the meteorologist indicated where severe storms might strike and indicated, from a weather sense, which stations were best situated to provide aircraft for convoy coverage. Additional services were begun for smaller navy vessels on patrol in the St. Lawrence River and Gulf in 1943. The next year, the British Admiralty asked for special daily forecasts for aircraft operating in the mid-Atlantic from ships without meteorological officers.

Pacific Coast

On the Pacific coast, the Vancouver Western Air Command forecast office undertook to provide services for the fleet operating out of Esquimalt. Fleet forecasts were prepared and issued twice daily and operational naval officers were briefed as required. Since the operations room was in the same building as the RCAF headquarters and the forecast centre, the naval duty officers had access to the same maps, hourly observations, and weather forecasts as did the RCAF officers. Prior to sailing, navy officers often spent time in the forecast centre where the meteorologists briefed them on weather conditions along the coast.

Naval Metmen

In addition to McElheran a number of other metmen resigned from the Meteorological Division to obtain navy commissions. These included J.H. Meek, the metman at Rockcliffe who began providing information to the navy without Patterson's permission; Dave Stewart and Bill Moore, metmen posted in early 1942 to RCAF flying training schools after the fourth short course; and G.S. Noel, who had been a navy cipher clerk. At the end of the war all except one commissioned meteorological officer resigned and arrangements were then made for the Meteorological Division to provide a meteorological service for the postwar Royal Canadian Navy.

CANADIAN ARMY

There was relatively little contact between the Meteorological Division and the Canadian Army during World War II. The Royal Canadian Artillery asked for and obtained upper air forecasts to aid in calibrating the guns at various training and operational sites throughout the country. At the request of the Department of National Defence, meteorologists and metmen were seconded to the army for secret chemical warfare tests at Suffield, Alberta. Also, metmen were seconded to the army to provide weather reports and forecasts for several "cold weather" exercises carried out at the end of the war.

Royal Canadian Artillery

As with the navy, the army followed the traditional practices and procedures of their United Kingdom opposite numbers and rarely sought meteorological information from the national weather service. There is no record of the prewar Canadian Army seeking or obtaining "meteor-telegrams" as the meteorological messages for the artillery were called. However, by the time the army did so, in November 1940, Patterson had already obtained from Britain a copy of the British Meteorological Office publication *The Supply of Meteor Reports to Artillery Units* and so was able to begin supplying them at once.

To correct the trajectory of shells through the atmosphere, artillerymen made certain corrections at the time of firing. The most important factor was the speed and direction of the upper winds at the levels through which the shell would pass. Temperature and humidity in the air strata are of less importance than the wind but all factors were grouped together into one correction called the "ballistic wind." Some meteorological offices made the calculations and sent the meteor-telegrams to artillery units to use but some of these units preferred to make their own calculations based on meteorological data sent from the meteorological office. Patterson objected to this since he feared mistakes would be made and the Meteorological Division blamed. But the calculations were time consuming so Patterson requested and obtained extra employees on the Pacific coast to make them in the Western Air Command forecast office. By October 1943 there were four Meteorological Assistants Grade 2 computing and sending eight meteor-telegrams a day to West Coast artillery sites from that office.

The Royal Canadian Artillery had several operational sites on the Atlantic and Pacific coasts. Tests and training exercises were conducted at Petawawa, Ontario, at Camp Shilo, Manitoba, and at Wainwright and Suffield in Alberta. The Meteorological Division prepared most meteor-telegrams but some units prepared their own corrections from the furnished meteorological data. Pilot balloon equipment was loaned to the army for use at the training sites so that the upper winds could be measured at the times of firing. Radiosonde equipment was provided to the Suffield station to allow the army to obtain additional upper air information.

Suffield

The most secret of any wartime establishments in Canada with which the Meteorological Division was associated was the Suffield Experimental Station, a forty-square mile area in southern Alberta established for experiments in gas warfare. In addition, elaborate investigations into the "anomalous dispersion of short waves" were made at that station. In May 1941, the Department of National Defence authorized positions for two meteorologists and two metmen for Suffield and these were filled by meteorologists G.C.W. Tait and Herb Chadburn and metmen Ross Armstrong and K.M. Korven. In 1944, Olie Johnson replaced Chadburn and later the next year Walter Halina and Alf Ingall joined the staff. The gas warfare research at Suffield was part of a very secret joint United Kingdom–Canada project and the meteorological personnel did not make reports on their work to the Meteorological Division.

Cold Weather Exercises

Towards the end of the war, the army, in conjunction with the air force, began to carry out experiments and tests on cold weather clothing and equipment. The probable weather was very significant in planning the times and places for the tests and during the exercises both forecasts and observations were important. Two of these tests were carried out in the winter of 1943–44. One was conducted in dry, cold weather near Prince Albert, Saskatchewan, where J.A. Ferris was seconded to the army for the testing period and one took place in cold, damp, wet weather at Cape Spear, near Torbay, Newfoundland, where Bev Cudbird was loaned to the army. Meteorological equipment was provided for both tests and a course in weather observing given to army personnel who would take the observations. The metmen supervised the observing, interpreted the forecasts for the army personnel, and provided other meteorological advice.

The Canadian Army conducted three relatively large cold weather exercises in the winters of 1944–45 and 1945–46. Exercise Eskimo was staged during the first winter between Prince Albert and Lac la Ronge, Saskatchewan, in cold, dry weather and Exercise Polar Bear between Williams Lake and Bella Coola, British Columbia in milder, damper conditions. About 1,200 men were used in each exercise to gain experience in moving men, equipment, and supplies in various weather conditions. Two metmen were provided for each exercise, one to remain at the base and provide forecasts and the other to accompany the expedition as the commander's meteorological advisor and to supervise the weather observations taken on the trip.

A postwar, smaller but more rigorous, cold weather exercise was carried out in the winter of 1945–46 to test equipment and techniques in relation to the part they might play in the civil evolution of the Canadian North. Known as Exercise Muskox, some twenty-five army "oversnow" vehicles and eighty men travelled 2,400 miles (3,800 km) from Churchill, Manitoba, to Edmonton, Alberta, via Victoria Island in the Arctic in February and March 1946. The RCAF assisted in this exercise by keeping the moving force supplied with provisions, fuel, and spare parts. The Meteorological Division provided a metman to accompany the expedition and set up, successively, short-term meteorological offices at Churchill, Norman Wells, and Fort Nelson as the expedition progressed. Weather reports and forecasts were transmitted by radio to the metman on the trip where he advised the commander and supervised the army personnel in keeping weather records. Gordon McKay and Don Storr were two metmen involved in Exercise Muskox.

RADAR DEVELOPMENT

In their experiments with radar at Ottawa the Canadian Army noticed echoes on the screen which appeared to come from clouds whenever rain was falling, and in June 1944 the army asked the Meteorological Division for assistance in studying the matter. The Division co-operated by providing weather reports and forecasts during the investigations and arranged for the army to telephone climatological station observers in the Ottawa area when there was a likelihood of rain. The observers, who were not told of the purpose of the calls, were asked about the cloudiness at the time and whether or not it was raining. Assisted by these visual observations the scientists began to be able to determine from the radar the area over which rain was falling and its direction of movement.

Additional research was conducted at the RCAF Clinton Radar School in the summer of 1945. A metman, Don Perrie, who had been at the Ottawa SFTS and had had some dealings with the radar scientists the previous year, was posted to Clinton and was in charge of the meteorological aspects of the work. The project was developed to the extent that Perrie was able at times to advise the Toronto Malton DAFO of the areal extent of precipitation during a storm.

WEATHER RECONNAISSANCE FLIGHTS

After the war began, the necessity for radio silence kept ships at sea from transmitting weather observations. To overcome this handicap for weather map analysis and forecasting in the British Meteorological Office the Royal Air Force soon organized reconnaissance flights over the Atlantic Ocean and the North Sea. Weather permitting, the outward flights were at low level with periodic observations and descents as close as possible to the ocean surface for more observations. Ascents were also made periodically to 15,000 feet and the return journey was at this level. The Germans also made reconnaissance flights to the west especially after bases in France and Norway became available to them in 1940. Later in the war the United States air forces had a considerable weather reconnaissance program.

In the spring of 1943, Rube Hornstein, officer-in-charge of the Halifax forecast office, conferred with the RCAF about the need for observations over the Atlantic to the southwest and south-southeast of Halifax. Authorization was received from National Defence headquarters in December and trial flights began that month. Daily flights began from Yarmouth in February 1944 and continued until the end of the war. Although strong representations were made for the flights to continue because of the meteorological need, the flights ceased in September 1945. There is no record of Meteorological Division personnel participating in the flights although the Division provided instruments.

On the Pacific coast, Western Air Command suggested weather reconnaissance flights over the Pacific Ocean as early as April 1943 but aircraft did not become available until a year later when the meteorological service was asked to provide the instruments and supervise their installation. Inevitable delays occurred and it was not until December 1944 that the first flights took place and January 1945 before Ventura aircraft flew regular daily flights from Patricia Bay. The observations were broadcast in code at first but in July it was decided that the transmissions could be in the clear. The flights were discontinued in mid-September 1945 through lack of staff and funds.

OTTAWA ROCKCLIFFE FORECAST OFFICE

The Ottawa headquarters of the Royal Canadian Air Force required a small meteorological office to serve the communications and transport flights operating out of Rockcliffe airport. An office was established in the spring of 1941 when metman Brock McElheran was posted there to provide meteorological support for the communications and transport flights from RCAF headquarters. When McElheran left for the navy, J.H. Meek was posted there a few weeks after completing a later short course. RCAF airmen began a full observing program at the airport in January 1942.

In 1942, because some of the flying from Rockcliffe was for long distances, the RCAF requested that the office be elevated to forecast centre status. There were more pressing needs for meteorologists and so the first meteorologists, Les Tibbles and Bob Dodds, were not posted to Rockcliffe until January 1943 when the office began a daytime forecast service. Meek resigned and joined the navy as a meteorological officer at about this time.

The office hours were extended in early 1944 when Graham Potter and Don McGeary arrived. The meteorologists prepared route forecasts and briefed aircrew for flights to other Canadian and American airports, locally and throughout the continent. These four meteorologists remained at Rockcliffe for the duration of the war.

WEATHERSHIPS

Although merchant and naval vessels had long taken weather observations from the oceans it was the twentieth century before the observations could be transmitted by wireless to shore stations for use in weather forecasting. During the flight of the British airship R-34 to North America in 1919, a number of ships were stationed along the route to transmit weather observations and to be available if an emergency arose. Similar measures were taken in 1930 when the R-100 crossed the Atlantic Ocean.

In July 1937, the French stationed the first weather ship in the Atlantic Ocean. The ship was stationed at 38 degrees North latitude and 44 degrees West longitude. Regular synoptic observations were taken as well as two radiosonde observations each day. The last cruise of the ship was disrupted by the outbreak of war in 1939.

There were weatherships in the Atlantic for various periods during the war but there is no record of Canadian participation. The Americans had weatherships in the Pacific Ocean but none were far enough north to be of value to Canada. The RCAF believed they were severely handicapped in their operations off the Pacific coast by the lack of weather reports and so, in July 1944, they asked for a weathership to be placed 500 miles off the coast. The Royal Canadian Navy agreed to supply a ship for this purpose but it was May 1945 before the HCMS *Woodstock* sailed from Vancouver. Requests for a second ship to allow one to be on station at all times were denied.

The Meteorological Division agreed to supply the observing staff, radiosonde instruments, and other supplies. Of concern was the fact that by using Division technicians there would be civilians aboard a naval ship in an area of operations. Having the men wear the uniform of the Marine Service of the Department of Transport solved this and agreement was obtained regarding the men's duties, responsibilities, and their relationship with the naval staff.

When Japan surrendered, the navy, at the urgent request of the RCAF and the Meteorological Division, agreed to continue to operate the ship as long as they could but a second ship was still not made available. In October 1945 the period of a cruise was changed from eight to eleven days and radar was tried on board to get upper winds but this attempt was only partially successful. With the navy running short of personnel the last cruise ended in mid-March 1946. It would be November 1947 before the Meteorological Division was again involved in weatherships and this was with Ocean Weather Station Baker in the Atlantic Ocean.

JAPANESE FIRE BALLOONS

In January 1945 a large air-borne paper balloon carrying instruments was shot down at Fort Simpson in the Northwest Territories. Over the next several weeks a large number of similar balloons were found on the ground in various parts of western Canada and the United States; one had travelled as far east as Michigan. The balloons were carried eastward in the prevailing westerly winds and were equipped with a pressure device that allowed the dropping of small bags of sand to enable it to maintain a constant elevation. After the altitude maintaining device was exhausted the balloon's mechanism was designed to drop its load of either a small explosive or an incendiary bomb.

The balloons caused little or no damage and all news of them was kept secret from the public at the time as any mention of them in the press or on radio was prohibited. Studies of the balloon problem were carried out in the United States and in Canada where the Canadian Army conducted investigations at Vancouver. There, at the Western Air Command forecast office, a meteorologist was assigned to assist by "back-tracking" the balloon from the point where it was found or sighted to a possible point of origin. Upper level pressure maps were drawn and studied for the North Pacific at 20,000 and 30,000-foot levels (6.1 and 9.1 kilometres). Speeds and directions were estimated at these heights from the maps in order to calculate the probable paths of travel of the balloons. All projected paths were found to cross the Japanese islands and after the war it was confirmed that the balloons had been released from a site near Tokyo. It was also revealed that the Japanese stopped the program when they received no indication that the balloons had ever reached North America.

MANUFACTURE OF INSTRUMENTS

From its establishment, the national meteorological service had manufactured and repaired meteorological observing instruments for its own use. John Patterson, an excellent instrument man, maintained a small but very efficient staff of instrument makers and assistants at Head Office. The Department of Munitions and Supply became aware of this early in the war and, in June 1940, asked him to have some work done on artillery gun sights that had proved to be very difficult for other instrument makers to do. Patterson arranged for his instrument makers to work on the instruments outside office hours at hourly rates and the work was done better and more quickly than had been thought possible.

Accordingly, the Department of Munitions and Supply requested that the Meteorological Division enlarge its instrument shop and have special machinery installed to undertake the manufacture of some specific secret instruments and to repair and recondition others. Besides the quality of the work the Division instrument makers could do there was the fact that no large commercial companies had been interested because mass production methods would not be viable when the number of instruments required was small.

The businessmen and engineers in Munitions and Supply wanted John Patterson to immediately contract for a new building and purchase the necessary machinery as a private contractor would do. This Patterson could not do as a civil servant but, after some delay, he received official approval from the Department of Transport late in December 1941. The building was completed in the spring of 1942, the procurement of the necessary machines and equipment was given priority, and a special ten-week course in fine instrument making was given at a nearby technical school to twenty-five or so young women. They were hired that spring and the Meteorological Division began to manufacture instruments. The chief instrument maker supervised the new staff and by the close of the war 3,050 instruments of twenty-one different types had been manufactured.

After the war the building and equipment were made available for purchase by the Meteorological Service which "entered the post-war period with one of the best equipped workshops for meteorological purposes anywhere in the world." [John Patterson in his manuscript "History of the Meteorological Service of Canada During World War II."]

[89] Brock McElheran, *V-bombs and Weathermaps: Reminiscences of World War II* (Montreal: McGill-Queen's Press, 1995).

CHAPTER 17

CIVIL OFFICES

Before the war, the various meteorological offices throughout the country with full-time employees, apart from those on the transcontinental airway, were unofficially called "Branch Offices." The term "outside offices" had been used earlier for these offices. In addition to the Branch Offices there were, of course, the synoptic weather observing stations where part-time observers were paid a nominal amount for taking two or three observations a day and the climatological stations where the observers were mostly co-operative volunteers without pay.

Victoria, British Columbia, was the only Branch Office where public weather forecasts were prepared and issued. With the organization of the Air Services Branch in the new Department of Transport in 1936 and the advent of Trans-Canada Airlines in 1937, the Meteorological Division began to organize and staff offices to provide observations and aviation forecasts for the pilots and airline dispatchers. These offices, staffed by meteorologists, meteorological assistants, and teletype operators were called District Aviation Forecast Offices.

BRANCH OFFICES

Victoria

The most important Branch Office was the one at Victoria Gonzales Heights where public weather forecasts were made and issued for regions in southern British Columbia. A long-time employee, F. Napier Denison, had been the forecaster for three decades before he retired in 1936. William Thorn moved from Toronto to replace him as the forecaster and he remained there for two years or so. Shortly after the Western Air Command forecast office opened at Victoria in 1940 the meteorologists in that office took over the British Columbia public weather forecasting.

The Victoria office, with a staff of five, provided a valuable link with the public in British Columbia by maintaining an archive of climate data and servicing requests for forecasts, climate data, and associated information pertaining to that province. Daily synoptic observations were taken and the office administered the climatological station network in the province. All the British Columbia stations sent their monthly reports to Victoria where they were processed and summarized before the reports were sent to Toronto for the national archive.

Vancouver

Known as the Vancouver Port Meteorological Office, this downtown office took full synoptic weather observations until March 1943. E.B. Shearman had replaced his brother, T.S.H. Shearman as observer in 1933. Prior to the war, Shearman visited with ships in the harbour to inspect their observing instruments and to encourage the ship's officers to transmit weather reports when at sea. From 1943 to 1945 Shearman maintained the "Port Met Office" and took climatological observations.

Edmonton and Moose Jaw

Two ladies staffed these one-person offices in Alberta and Saskatchewan, Mrs. Edna Owen in Edmonton and Mrs. Edna W. Ogilvy in Moose Jaw. They were sometimes called "provincial agents" and monitored the climatological stations in their respective provinces as well as taking daily synoptic observations and telegraphing them to Toronto. When observing stations were opened at the airports their observations were no longer needed but Mrs. Owen kept reporting until 1943 and Mrs. Ogilvy until 1948. Their prime responsibility was, however, to collect monthly report forms from the climatological observers and perform some clerical processing on the data before sending the forms to Toronto. They had only minor contact with the public but kept the climatological station observers provided with report forms, envelopes, and other stationery and referred any observer problems to Head Office.

Winnipeg

The Winnipeg Grain Exchange office had been opened in 1920 to provide a specialized forecast and climate services for the grain trade in the Prairie provinces. The Winnipeg airport office had taken over the synoptic observing and the Grain Exchange employees ceased taking observations at St. Johns College in 1938. The establishment called for three employees and the officer-in-charge was meteorologist Hal Troop.

A weather map was drawn and published each day and, every week in the growing season, a *Weather Summary for the Prairie Province* was prepared and issued. To provide up-to-date data for the *Weather Summary* a few score climatological station observers across the prairies were each paid a nominal sum to send postcards to Winnipeg at the end of every week giving their daily temperature extremes and precipitation.

Quebec City

The Quebec City Observatory was one of the first telegraph reporting stations when the Meteorological Service was started in 1871. Maurice J. Royer had been named the observer in 1929 and the original Quebec observatory building was demolished in 1936. Royer continued to take and transmit weather observations each day during the war and provided a climate and forecast information service to the public.

Saint John

This New Brunswick observatory also dates back to the Confederation years and the beginnings of the Meteorological Service. Francis M.

Barnes had become the observer in 1927 and there was also a stenographer and a caretaker on staff. During the war years, Barnes continued to take three synoptic observations a day and collected the climatological report forms from observers in the Maritime provinces each month. Besides processing the data and sending the reports to Toronto he maintained a clerical archive of climate data for the region and provided a climate information service to the public.

Other Offices

There were a few other civil meteorological offices in wartime but they were not classed as Branch Offices. One was at the Montreal McGill University Observatory, which had been one of the earliest synoptic reporting stations in the country. The McGill observer was never a Meteorological Service employee but the university was given an annual grant as a contribution to the cost of the observatory. Synoptic observations continued to be taken at the observatory in wartime where the observer maintained an archive of climate data and provided information to the public when asked.

Two minor observatories continued to exist in wartime at Meanook, Alberta, and Agincourt, Ontario. Their prime function was to take magnetic observations. Herbert E. Cook became the Meanook observer in 1916 and took a full set of meteorological observations until 1940 but he did not telegraph them to Toronto. During wartime F. Furnell was the observer at Agincourt, a small observatory just northeast of Toronto, where magnetic observing had been moved in 1898 to escape the effect of the streetcar system on the Toronto observations. Only climatological observations were taken at Agincourt during wartime.

AVIATION FORECAST OFFICES

There were six District Aviation Forecast Offices (DAFOs) located across Canada in wartime. They were on the transcontinental airway from Vancouver, British Columbia, to Moncton, New Brunswick, and existed primarily to serve Trans-Canada Airlines (TCA), now Air Canada. Airport DAFOs opened at Vancouver Sea Island and Winnipeg Stevenson Field in October 1937 and gave twenty-four-hour service to TCA beginning in February 1938. TCA service was extended to Montreal St. Hubert airport and Toronto Malton airport in June 1938 and partial service was commenced at those DAFOs. The Montreal DAFO began to provide around-the-clock service in April 1939 and Toronto Malton by that summer.

TCA service was extended to Calgary and Edmonton from Lethbridge on the transcontinental route in April 1939. In December 1940 Lethbridge became the TCA dispatching centre for western Canada and a DAFO was established there. TCA also extended their route to Moncton in the fall of 1939 and on to Halifax in 1940. Forecasts and briefings for the Montreal to the Maritimes route were done by the Montreal DAFO until the sixth DAFO could be established at Moncton in May 1942. The delays in opening the latter two offices and additional main forecast offices at other locations in the early war years were caused by the shortage of trained meteorologists.

Of the half dozen meteorologists hired before graduates of the University of Toronto course in physics (meteorology) first became available in 1934, only two became DAFO forecasters (McCauley and Chisholm). Hiring of the MA graduates for posting to the DAFOs began in the summer of 1936 and it was from these graduates that the new DAFOs were staffed, first Vancouver and Winnipeg and then Montreal and Toronto. (Original plans had called for a DAFO at Kapuskasing, a mid-point on the Winnipeg to Montreal airway, rather than at Toronto.)

By the end of 1939 the four original DAFOs were open twenty-four hours a day. Each had a staff of five or six meteorologists and the necessary assistants and teletype operators. Route and terminal forecasts, valid for eight hours, were issued four times a day. With the war, TCA flying increased markedly and added considerably to the meteorologists' workload since individual forecasts, strip maps, and personal briefings were given to each flight before departure. This workload was increased further by May 1940 when flying training began at RCAF airfields where metmen (dependent forecasters) needed the RCAF analyses and forecasts issued by the DAFO meteorologists for guidance. Also, as the new graduates were hired on completion of the MA courses they were in demand at the RCAF operational stations as well as at the new DAFOs. Most forecast offices were markedly understaffed until the situation began to improve when the first advanced course graduates became available in 1942.

In reviewing the rosters at the various DAFOs in the early war years, one notes immediately how experienced meteorologists at the first four DAFOs were moved to be the nuclei of the new DAFOs and other forecast offices that opened later. In September 1939, the Montreal and Toronto DAFOs each lost two meteorologists, half their experienced staff, to the new Halifax Eastern Air Command forecast office. The same thing took place a year later on the Pacific coast where Vancouver lost two experienced men to the new Western Air Command office and on the Prairies when both Vancouver and Winnipeg contributed meteorologists to the new Lethbridge DAFO. And later, in 1942, meteorologists from the Toronto and Montreal DAFOs were moved to the new Moncton office to begin providing service there.

Vancouver

Aviation weather observations were commenced at Sea Island airport in September 1936. In June 1937 Alan McCauley arrived to set up the first District Aviation Forecast Office and in September began to provide a partial service for TCA flights to Seattle. McCauley had spent a short period with Archibald in Winnipeg before he went to Vancouver. Over the next year he was joined by Carl Mushkat in December, Hank Edwards in February 1938, George Pincock in May 1938, and Al Crocker in July 1938. These five meteorologists provided forecasts for the Vancouver to Winnipeg route when TCA scheduled flying began that summer.

Don McIntyre joined the staff in March 1940 but departed that summer to organize a Western Air Command forecast office for the RCAF in Victoria. In the fall of 1940 Vancouver lost three more experienced forecasters — Mushkat and Crocker to the new Lethbridge DAFO and Pincock to head the new Victoria office. Clarence Cross arrived at Vancouver that fall but it was well into 1941 before Don Cameron and John MacNeil came. The staff was not significantly augmented until the graduates of the advanced courses became available. Then, Dave Strachan and Jack Wright came in 1942, Henry Bellhouse and E.N. Ellis in 1943, and Jack Henderson and Ken Harry in 1944. Each of these meteorologists remained at that DAFO for the duration;

Wright and Strachan remained until retirement decades later.

The service records of meteorologists who left the Meteorological Division after the war are not available but it is known that MA graduate Fred Kelly forecasted at Vancouver in the early war years and such meteorologists as Bill Cameron, Art Wright, and H.S. Keenleyside were there for parts of the 1943–1945 period. The Vancouver DAFO was relieved of forecasting and briefing responsibilities for the RCAF when the Victoria WAC office commenced operations. It was, however, responsible for an area of the country where forecasting was made difficult by the lack of observations from the Pacific Ocean. Further, the office was responsible for forecasting for flights along the airway through the mountains to the Prairies, probably the most difficult TCA route for which to forecast.

Winnipeg

In 1937 Don Archibald returned to Winnipeg where he had established and supervised the Western Air Mail Weather Service for two years before its demise in 1932. Dean Smith joined him late in 1937 and Des Kennedy and Hal Troop early in 1938 to provide partial service to TCA. Two left in 1939, however, Troop to the Winnipeg Grain Exchange office and, at the beginning of the war, Des Kennedy to Head Office to develop observing and instruction manuals and to train recruits. In the summer of 1938, Sam Neamtan and Mel Robertson had arrived allowing the office to remain open twenty-four hours a day. G.C.W. Tait came in early 1939 and Frank Benum in July of that year.

Don Archibald had supervisory responsibilities for aviation meteorology service in western Canada and organized the establishment of the Lethbridge DAFO in December 1940. Dean Smith, Sam Neamtan, and Frank Benum left Winnipeg to staff the new office and H.L. Osmond was their only replacement leaving the office significantly short-handed for some time. When Tait left for Suffield in mid-1941 Clarence Penner replaced him for a few months and then Burn Lowe replaced Penner. Syd Buckler came in September 1941 and staff strength remained at five until Jack Labelle came early in 1944 and Hugh Cameron later that year.

Late in 1945 the meteorologist strength was bolstered when experienced meteorologists became available as the wartime forecast offices were closed. Those arriving at that time included Henry Capelle, Van Gordon, Einar Einarsson, and Ralph Anderson.

Until a forecast office was established at the Rivers RCAF station in early 1943 the Winnipeg DAFO issued regional RCAF forecasts to provide guidance to the metmen at the many RCAF training stations that had opened on the Prairies. Metmen at several stations complained that the morning forecasts were never received in time to be useful and that they were not detailed enough to be of value. It is not known why the RCAF forecast could not have scheduled for an earlier release from the Winnipeg DAFO but it is apparent that the office was notably understaffed in those early war years and their area of responsibility was large.

Toronto

It is interesting to note that in planning for meteorological services for the eastern TCA route, a District Aviation Forecast Office was planned for Kapuskasing airport and not for Toronto. With the limited range of aircraft (Lockheed Electras and later Lodestars) then in use it was thought that there would be regular landings at Kapuskasing, the midpoint on the Winnipeg to Montreal route, and that crew briefings would be necessary there. When TCA began to provide partial service at Toronto in October 1938 Patterson did not have a sufficient number of meteorologists available to staff a separate office at Malton airport so meteorologists at Head Office took the shifts necessary to keep the office open. Fred Turnbull was named the officer-in-charge and by the summer of 1939, when TCA began full service, meteorologists Alex Chisholm, Don Ross, Jim Leaver, and Harvey Halbert were assigned full-time to the Malton DAFO.

The opening of the Halifax Eastern Air Command forecast office in September depleted the Toronto Malton office as Ross and Halbert were posted to the new office. When Leaver was moved to the Montreal St. Hubert DAFO in April 1940, help was again required from Head Office until Frank Thompson and Bill Rae, 1939-40 MA course graduates, were posted to Malton. Bob Graham arrived a year later from the 1940-41 course and the staff remained relatively stable for the rest of the war. Peter Sandiford arrived from the first advanced course in February 1942 but resigned early in 1943 and was replaced by Duff Butterill from the third advanced course. Gil Clark and Archie McCracken came in April when the short-lived Head Office RCAF forecast office was closed. Clark went on to other postings and McCracken resigned in 1945 when several others came to the Malton staff including Paul Denison, Norm Jefferson, Mac Elsley, Rollie Kendall, Don Boyd, and Herb Chadburn.

In addition to issuing the TCA route and terminal forecasts, the Malton office was responsible for the RCAF area forecasts that provided guidelines for the many RCAF flying training schools in southern Ontario. (The special forecast office set up at Head Office in 1943 for this purpose lasted but a few months.) Because of the extensive wartime industrial work in Toronto and the introduction of TCA routes to New York City and other American cities there was probably more civilian flying requiring forecasts and briefing at Toronto than at any other Canadian airport in wartime. For several months early in the war, Turnbull and some of his staff also looked after the meteorological lecturing and forecasting duties at the first Air Observers School at the airport before a metman was posted to the school.

Montreal

Carl Mushkat was the first meteorologist at St. Hubert airport where he worked at developing transatlantic forecasting in 1936 and 1937. Fred Mahaffy came to St. Hubert to organize a DAFO in the summer of 1937 and Mushkat was posted to Vancouver. TCA began partial service to Montreal in June 1938 and by August a full scheduled service was flown. In September 1941 all civil aviation and the DAFO moved from St. Hubert to the new Dorval airport.

Reg Noble (officer-in-charge), Fred Mahaffy, Ralph Anderson, Ted Wiacek, Morley Oretzski, and Rube Hornstein first staffed this DAFO. With the outbreak of war in September 1939, Noble and Wiacek left to staff the new Halifax forecast office. The Montreal office was further depleted when Hornstein was posted to Halifax in March 1940 to succeed Noble as the officer-in-charge. Jim Leaver replaced him and, in turn, Leaver was replaced by Don Ross who remained until the summer of 1941.

Clarence Penner and H.L. Osmond came to St. Hubert from the 1940 MA course; Osmond remained for a few months and Penner for a year. In late 1940 Hank Sabraw replaced Ralph Anderson and Harry Greb was on staff for three months in 1942–43. Douglas Chapman, Norm Jefferson, and Al Mason arrived in late 1942 and the composition of the office remained fairly stable until 1944 when Paul Johns and Don McClellan arrived. Several meteorologists who resigned after the war, such as I. Sheichet and L.R. Mumford, were also at Dorval for brief periods in 1944 and 1945.

Moncton

The Moncton DAFO began operations in mid-1942 and took over from Montreal responsibility for TCA forecasting in the Maritimes. Earlier, late in 1939, when TCA began to fly to Moncton, meteorologist Hank Sabraw had been posted to the meteorological office at the Moncton airport to advise and brief pilots but he did not have the necessary staff and facilities for map analysis and forecasting.

Hourly observations were taken at the airport by Meteorological Division observers and a Service Flying Training School was established there in December 1940. The school operated at the airport until January 1944 and was closer to the flying field than the DAFO. For this reason many itinerant pilots went to the school meteorological office for briefings rather than to the DAFO. It was hard to avoid this so the metmen received daily briefings from the DAFO meteorologists to prepare them to give such briefings.

Morley Oretzski was posted from the Montreal DAFO to organize and head the new Moncton office in mid-1942. Bill Rae and Frank Thompson came from the Toronto DAFO and Don Day from the second advanced course to complete the original staff. Within a few months, John MacNeill and Les MacHattie, both from RCAF operational station postings, came to replace them. Frances Carson Sutherland and William Buscombe, new advanced course graduates, were on staff for parts of 1944 and 1945 before Sutherland resigned and Buscombe left for the University of Saskatchewan. Reid Dexter, Merv Fleming, Larry Cooke, and Conn Sutherland joined the staff before the end of 1945 to replace those who had resigned.

Lethbridge

Kenyon Field, Lethbridge's airport, was constructed in 1938 and

Trans-Canada Airlines began to use it for passenger service in April 1939. The Lethbridge airfield was important in the early days of TCA since it was the point of departure for flights from the Prairies over the Rocky Mountains to Vancouver and the Pacific coast, and a terminal for flights to Calgary and Edmonton.

It was December 1940 before there were sufficient meteorologists available to establish a Lethbridge District Aviation Forecast Office. That month, Dean Smith and Frank Benum arrived from Winnipeg, and Al Crocker and Carl Mushkat from Vancouver, and airways forecasting commenced. Meteorologists Herb Chadburn, George Legg, and Olie Johnson were at Lethbridge for brief periods in 1941 and 1942. Tommy Thompson came in 1941 and in 1942 Hugh Cameron replaced Al Crocker who was moved to Toronto to assist in training metmen and meteorologists. Horace Wilson, Sam Neamtan, and Charlie Goodbrand came in 1943 and 1944. Late in 1945, when other forecasting offices were closing, Gordon Muttit, Ray Walkden and Erling Anderson joined the staff.

The Department of Transport continued to operate the airport during the war years although two RCAF schools under the BCATP, an Elementary Flying Training School from 1940 to 1941 and a Bombing and Gunnery School from 1941 to 1944, used the field along with the USAAF and TCA. The importance of the Lethbridge airport was diminished a few years after the war when TCA began to fly from Winnipeg to Vancouver through Calgary.

Other Civil Forecast Offices

As described in other chapters, full service forecast offices existed at Gander, Goose Bay and a second office at Montreal Dorval for Ferry Command and other transatlantic flying; and at Halifax and Victoria/Vancouver for RCAF operations. (The Goose Bay and Gander offices served both Ferry Command and the RCAF: both offices were split into two offices for a period in 1944.) Other forecast offices at Edmonton, Whitehorse, and Prince George existed for military flying, as did the RCAF offices at Rivers and Ottawa Rockcliffe that have been described elsewhere.

Although never staffed to the extent that full service could be provided, some intermediate airports on the transcontinental airway required meteorologists in the early war years. Mention has been made of Sabraw at Moncton before the DAFO was opened and Clarence Cross was at Ottawa Rockcliffe for several months in late 1940. Military flying required a meteorologist (Lloyd Richards) at Kapuskasing for eighteen months in 1943 and 1944. Late in the war when peacetime aviation needs were uncertain, Erling Anderson and Ralph Anderson were at the Regina office for two or three months.

CHAPTER 18

OBSERVATIONS

In the 1930s the Meteorological Division established two dozen new telegraphic and wireless synoptic weather reporting stations in the northern portions of the provinces from British Columbia eastwards to Labrador and two stations on the Arctic islands. The only operational or "real time" stations in the North before that decade were the dozen or so Royal Canadian Corps of Signals stations in the Yukon and the Mackenzie River basin and four Marine Radio stations along Hudson Bay and Strait.

The expansion of the observing network in the North during the late 1930s was accompanied by a significant expansion in the South along the new transcontinental airway. This was made possible by resources allocated to the Radio and Meteorological Divisions to enable them to install and provide the necessary air navigation services for the new Trans-Canada Airlines. Observations were taken by radio operators at radio range stations and by meteorological observers at terminal and intermediate airports.

With these expansions, the number of synoptic weather reporting stations in the country increased from sixty in 1930 to seventy-seven in 1934 and to 136 in 1939 when the war began. At that time, thirty-one stations, mostly at airports, were transmitting synoptic and hourly observations on the meteorological teletype system. Forty-five stations in the North and on the coasts were reporting by wireless, and sixty traditional telegraph reporting stations were sending in observations, mostly twice daily.

In wartime, the Meteorological Division established only a few new stations, mostly in isolated areas while, during the same period, the total Canadian observing network was expanded greatly as many stations were established by the Royal Canadian Air Force and the United States Army Air Force. All RCAF operational stations had meteorological sections where airmen took the observations as they did at the air training stations, except at those stations where flying was local and there was little need for a meteorological section.

Under the North American defence agreement, the USAAF established about four dozen observing stations in the northwest part of the country where there was considerable flying along the Northwest Staging Route and the Mackenzie River airway. A smaller number of stations were established in northeastern Canada to supplement the Canadian observing network along the Crimson or Northeast Ferry Route.

When observations were commenced at airports along the transcontinental airway just before the war, several of the traditional telegraph reporting stations became redundant and some of them were closed in wartime. It was a similar situation when observations were commenced at some RCAF airfields, which led to the closing of some nearby town telegraph reporting stations. After the war only a fraction of the several dozen RCAF observing stations remained open by the end of 1945. Most USAAF stations closed in 1945 although several observing stations at airports in northeastern Canada and along the Mackenzie River airway were taken over by the Meteorological Division.

The opening and closing dates for Meteorological Division observing stations can usually be found fairly accurately, but that is not the case at stations manned by the RCAF, the USAAF, and other agencies. Precise opening and closing dates for the air training schools have been published but at some schools the meteorological observing program did not start for some weeks and at others observations continued for weeks after the school closed. All RCAF observing stations were mandated to send the standard Meteorological Division observation forms to Toronto at the end of each month for archiving but a few failed to do so.

After the war the United States Army Air Force undertook to send copies or microfilm copies of the observations taken in Canada in wartime to the Meteorological Division for archiving. Inspection of these data sheets indicates the approximate dates when observations were started and ended. However, it is apparent that at some stations observations were commenced perhaps months before any forms or film were sent to Toronto. The United States Navy did not begin to send copies of their wartime Argentia observations to Toronto until March 1945.

Accordingly, some of the dates given in the paragraphs that follow for the opening and closing of specific RCAF and USAAF stations may be in error.

ATLANTIC CANADA

Except for Labrador, there had been a sufficient number of synoptic observing stations in each of the four Atlantic provinces for public weather forecasting. But, with the outbreak of war, more stations were needed for aviation meteorology and mention has already been made in an earlier chapter of the number of new observing stations in Atlantic Canada. Both the Meteorological Division and the RCAF opened new airport stations and in time most of the old part-time telegraph reporting stations were phased out. A brief history of the openings and closings of the wartime observing stations in each province is given below.

Nova Scotia

Before the war, observations were obtained from Halifax, Sydney and Yarmouth by telegraph, and from Sable Island and St. Paul Island by wireless. In September 1939, the Meteorological Division began taking four synoptics a day at Halifax and opened new stations in July 1940 at Ecum Secum and Liverpool. To start an observing program, a civilian and a few airmen observers were sent to the RCAF seaplane base at North Sydney in November 1939. At Halifax, the meteorologists supervised airmen observers at the Dartmouth RCAF station, where the first observations had been taken in 1938, in expanding and improving their observing program. RCAF airmen began to take airport observations at Yarmouth in February 1940, at Debert in January 1941, at Sydney in April 1941, and at Greenwood in May 1942. In the summer of 1945, observations were commenced at a new radio range station at Copper Lake and by Maritime Central Airways at New Glasgow. The RCAF took some observations at Shelburne in the spring of 1943 when a squadron was based there. Part-time observers at the Sydney and Yarmouth telegraph reporting stations ceased observing in March 1941.

On Sable Island, Transport's Radio Division maintained a wireless station where the radio operators had been taking synoptic observations since 1891 and pilot balloon observations since about 1930. When the war began in September 1939 codebooks were sent to the island in a few weeks so observations could be resumed. With the RCAF requiring forecasts for their patrols over the Atlantic, the Sable Island observations became very important to the new Halifax forecast office and to the RCAF in planning operations. In mid-1941 the RCAF requested hourly observations from the island; there was no accommodation for additional personnel so the existing radio staff agreed to take observations every two hours during daylight hours for an extra allowance. Later, in 1944, when more accommodation was built, the Meteorological Division installed a radiosonde station on the island.

Prince Edward Island

Synoptic observations were taken at the Charlottetown Experimental Farm before the war and the program was continued. At the Charlottetown airport, RAF airmen began observing in January 1942, RCAF airmen replaced them in 1944, and late in 1945 the observing duties were taken over by radio range operators. The RCAF began an observing program at the Summerside airport in March 1941 although a training school did not open there until May 1942.

New Brunswick

The prewar synoptic observations were taken in New Brunswick at Chatham and Kedgwick by traditional part-time observers and at

Saint John and Shediac by Meteorological Division employees. In October 1939, the Shediac station was closed, observers were posted to the Moncton and Saint John airports to begin observing programs, and radio range operators began to report from the Blissville airport on the transcontinental airway. The RCAF began airport observations at Pennfield Ridge in September 1941 and at Chatham in January 1943. (A training school had opened at the latter airport in July 1941 but sometimes there was no immediate demand from the Meteorological Division for the transmission of observations when there was a nearby telegraph reporting station.) The Kedgwick station closed in April 1943 but the telegraph reporting stations at Saint John and Chatham continued to report until after the war. The Meteorological Division opened a station at Campbellton in the summer of 1945.

Newfoundland and Labrador

When the war began there were ten stations in Newfoundland and Labrador. Meteorological Division staff took the observations at Newfoundland airport (Gander) and at the Botwood seaplane base. Marine Radio or Canadian Marconi operators took the observations at Belle Isle, at Cape Race, and at Cartwright in Labrador. Employees of Newfoundland Postal Telegraphs took the observations at Fogo, Grand Bank, Millertown, and St. George's. Observations were taken at Memorial College in St. John's until September 1941.

During the war the Meteorological Division began observations at Goose Bay in December 1941 and at Sandgirt Lake in Labrador in April 1942. The RCAF began observations at St. John's Torbay airport in January 1942 and took observations at Canada Bay during the latter half of 1942. Observations were received from the RCAF radar station at Cape Bauld, near the Strait of Belle Isle, from August 1943 to October 1944. Marconi wireless station staff began observing at Hopedale in January 1942; radio range staff began observing at Buchans airport in November 1943 and at St. Andrews, near Port aux Basques, in December 1943. Postal Telegraphs staff began observing at St. Anthony in September 1945.

The United States Army Air Force began observations at their Stephenville base in February 1942 and the United States Navy at their Argentia base at about the same time. The USAAF also began observing at Wesleyville and St. John's in 1943 and at Cape Harrison and Hebron in Labrador the same year.

QUEBEC

In 1939 there were only two synoptic observing stations reporting by wireless from the vast area of northern Quebec. One was a trading post at Fort McKenzie, 100 miles (160m kilometres) south of Ungava Bay, and the other was the Cape Hope's Advance Marine Radio station on Hudson Strait where observations ceased for two years after the war began. Much further south in the interior there was a wireless reporting station at Chibougamau and another at St.-Félicien, on the north side of Lac Saint Jean. Two prewar stations, Doucet and Sanmaur on the northern Canadian National Railways line, closed in late 1939.

In southern Quebec, in 1939, there were traditional synoptic reporting stations along the St. Lawrence from Montreal to the gulf — the McGill University observatory, the Quebec City observatory, Father Point, Clarke City, Fame Point, Anticosti (Southwest Point), and Harrington Harbour. Synoptic and airway hourly observations were taken at Montreal St. Hubert and Megantic airports.

The war brought new RCAF airport station observations at St. Jean d'Iberville in 1941, at Mont Joli and Bagotville (originally called Saguenay) in 1942, and at Quebec Ancienne Lorette in 1943. (A training school had opened at the Quebec airport in September 1941.) Observations were also commenced at the new Montreal Dorval airport in August 1941 and at Seven Islands (Sept-Îles) in 1944. The RCAF took observations at their Gaspé seaplane base during the summer and fall flying seasons of 1940, 1942, and 1943. In northern Quebec the USAAF began taking observations at Fort Chimo late in 1941, at Indian House Lake in 1944, and at Mingan and Mecatina along the airway from New England to Goose Bay in 1943.

Establishing observing stations in north and central Quebec was given high priority by the Meteorological Division early in the war. New stations in the interior were installed at Senneterre in 1940, at Nitchequon, Lake Manuan, and Lake Norman in 1942, and at Parent in 1943. On Hudson Bay, Great Whale River (Kuujjuarapik) and Port

Harrison (Inukjuak) began to report by wireless in 1942 and, on the Gulf of St. Lawrence, Natashquan and Baie Comeau (Outardes Falls) began to report in 1942 and 1943 respectively.

Weather observations were also taken on Mount Jacques Cartier in the centre of the Gaspé peninsula where the air force established a frequency-modulated-radio teletype station to be a connecting link between Gaspé and Sept-Îles. In the fall of 1943 RCAF airmen began to take observations and sent them to Gaspé on a local radio circuit. But, within days, a severe storm with high winds and icing damaged the instruments beyond repair. It was not until October 1944 that the program could be resumed and the observations sent out on a low powered radio circuit. In June 1945 hourly reports were commenced at the request of the USAAF and when this requirement ceased the RCAF closed the station early in November 1945.

ONTARIO

In 1939 there were four observing stations north of the transcontinental airway in northern Ontario. Observations were taken by Hudson's Bay Company (HBC) personnel at Trout Lake, by Canadian Pacific Airlines staff at Pickle Lake and Red Lake, and by a part-time observer at Moosonee. The only new station established in wartime north of the transcontinental airway was Lansdowne House where HBC staff began observing early in 1941. On the airway teletype circuit across northern Ontario there were observing stations at Kenora, Sioux Lookout, Wagaming (Armstrong), Nakina, Pagwa, Kapuskasing, Porquis Junction, Earlton, and North Bay. South of the airway traditional telegraph reporting stations with part-time observers existed at Port Arthur, White River, and Haileybury. The only new stations in wartime south of the airway were a new radio range station at Fort William in 1941 and a contract station at Cartier, near Sudbury, in 1945.

The prewar traditional synoptic observing network in southern Ontario consisted of stations at London (Lambeth airport), Southampton, Kingston (Marine) and Parry Sound with part-time observers and at the Ottawa Experimental Farm. Just before and during the first year or so of the war, civil airport and/or radio range station observations were commenced at Clear Creek, Windsor, London (Crumlin), Toronto Malton, Reay (Muskoka), Killaloe, Stirling, and Ottawa Uplands. The RCAF stations at Camp Borden and Trenton had commenced observations before the war began. The forestry research station at Chalk River took synoptic observations in summer during the war years. When observations were commenced at new nearby civil or RCAF airports, the part-time observers at Kingston Marine, London Lambeth, and Port Arthur ceased observing.

The station network in southwestern Ontario during the war years was probably the most dense of any network at any time in the country as more observing programs were commenced at air force flying training stations. Observations began at Aylmer, Brantford, Dunnville, Hagersville, Hamilton Mount Hope, and Port Albert in 1941; at Centralia and Ottawa Rockcliffe in 1942; at Mohawk (Deseronto) and Kingston in 1943; and at Pendleton in 1945. Hourly reports were taken and transmitted on the teletype circuits by many of these stations around the clock which enabled metmen at stations in the region to maintain a close watch on the development of synoptic weather situations and the resulting weather changes. Only a few of the RCAF stations in southern Ontario took synoptic observations.

Complete weather observations were continued at the Meteorological Division's Head Office in Toronto during the war years.

MANITOBA

In 1939, in northern and central Manitoba, there were synoptic observing stations reporting by telegraph from Churchill and The Pas and three reporting by wireless from God's Lake (Canadian Airways), Norway House (Hudson's Bay Co.), and Little Grand Rapids. In the southern part of the province there were stations at the Winnipeg and Rivers airports on the transcontinental airway and a traditional part-time telegraph station observer at Minnedosa.

Both God's Lake and Little Grand Rapids stopped reporting in 1944 but twelve new stations, mostly at RCAF and USAAF bases, began to report in wartime. Observations were commenced at RCAF flying training stations at Brandon, Carberry, and Portage la Prairie in 1941, at Dauphin in 1942, at Gimli and Souris in 1943, and at Paulson in

1945. (A Bombing and Gunnery School had opened at Paulson in 1941 but a meteorological section was not required until 1945.) The Department of Transport established a new radio range station at Neepawa in 1945. The USAAF established stations at The Pas and Churchill airports in 1943; on the Canadian National Railway's Hudson Bay branch at Gillam and Wabowden and at Duck Lake to the northwest of Churchill the same year. In 1945 and 1946, when the Americans departed, these USAAF observing stations, except the one at Duck Lake, were taken over by the Meteorological Division.

SASKATCHEWAN

There were no reporting stations in Saskatchewan north of Prince Albert in 1939. Traditional telegraph reporting stations in the province were located at Battleford, Moose Jaw, Prince Albert, and Qu'Appelle and airways stations on the trans-Canada teletype circuit were in place at Swift Current, Regina, and Broadview airports. The telegraph reporting stations at Battleford, Prince Albert, and Qu'Appelle ceased reporting in wartime during which six new airport stations began to report. The Moose Jaw, Saskatoon, and Yorkton air force stations began reporting in 1941; the North Battleford station in 1942; Weyburn in 1943; and Dafoe and Estevan in 1944.

The USAAF established observing stations at Hudson Bay Junction in 1943 and at Island Falls in 1945. Both were taken over by the Meteorological Division late in 1945.

ALBERTA

There were ten reporting stations in Alberta when the war began. Northern Alberta was much better covered than the northern portions of the other two Prairie provinces with a station at Keg River reporting by wireless and Fort McMurray, Beaverlodge, and Fairview by telegraph. Other traditional telegraph reporting stations with part-time observers were Banff, Jasper, and Edmonton (city). The airport stations on the transcontinental airway and the airway to Edmonton in 1939 were Cowley, Lethbridge, Medicine Hat, Calgary, Penhold (Red Deer), and Edmonton.

Nearly twenty new observing stations appeared in wartime. The RCAF began reporting from Calgary Currie Field and Claresholm in 1941, from Macleod and Pearce in 1942, and from Vulcan in 1943. The Meteorological Division opened stations on the approach to the Northwest Staging Route in 1942 at Grande Prairie and Whitecourt and others in 1944 at Coronation and Lac la Biche. In 1945 radio range operators began taking observations at Vermilion.

Since the airway to Alaska crossed Alberta the USAAF supplemented the Canadian network of observing stations by establishing stations at Wagner in 1943 and at Fawcett, Edmonton Namao airport, Peace River, and Valleyview in 1944. On the Mackenzie River airway there were Royal Canadian Corps of Signals stations reporting twice daily synoptic observations in 1941 but the Americans needed hourly reports from the airports they were building. Accordingly the USAAF began reporting hourly observations from Fort McMurray in January 1944 and opened a new station at Embarras early in 1943.

BRITISH COLUMBIA

Except along the airway to Alberta, the density of the meteorological observing network in British Columbia before the war was insufficient for proper synoptic map analysis and forecasting. Meteorological Division observers or radio range operators at Vancouver, Princeton, Carmi, Crescent Valley, and Cranbrook reported hourly on the teletype while Copper Mountain, Grand Forks, and Hope were telegraph reporting stations near the airway. Vancouver city and Victoria Gonzales Heights observatory continued to take synoptic observations. There were Marine Radio stations along the coast where two observations a day were taken — Alert Bay, Dead Tree Point, Estevan Point, Kains Island, Langara, and Pachena Point. In the interior, telegraph communications were available for observers to report twice daily from Atlin, Prince George, Kamloops, Prince Rupert, and Williams Lake while the observers at Fort Nelson and Hudson Hope used radio facilities.

The RCAF established a meteorological section at each of the new seaplane bases on the coast in 1941 and 1942 — Alliford Bay, Bella

Bella, Coal Harbour, Ucluelet, and Prince Rupert. Observations from RCAF airfields began in 1940 at Patricia Bay, in 1943 at Boundary Bay, in 1942 at Tofino, in 1944 at Abbotsford, Comox and Port Hardy, and in 1945 at Sandspit on the Queen Charlotte Islands. Along the coast, new Marine Radio stations were put in place in 1944 at Bull Harbour and Cape St. James. By 1945, there was sufficient staff at the Marine Radio stations on the coast to schedule four synoptic observations a day. Also, Transport established a radio range station in 1944 on the Queen Charlotte Islands at Massett where observations were taken.

On the Interior Staging Route, Transport established a radio range station in 1942 at Smithers and stations at Ashcroft, Dog Creek, Terrace, and Vanderhoof in 1944 where the civilian Radio Division operators were responsible for four synoptic observations a day and hourly observations. Meteorological Division observers took the observations at Prince George airport beginning in 1942, and along that portion of the Northwest Staging Route in British Columbia at Beatton River and Fort St. John beginning in 1942, and at Smith River in 1944. During the war the USAAF began to take observations at locations along or near the staging route at Coal River in 1942; at Camp Blueberry, Dawson Creek, Log Cabin, Muncho Lake, Summit Lake, Trout Liard, and Trutch in 1944; at Morley River in 1945. Other USAAF observing stations in the province were at Dease Lake, Nelson Forks, and Takla Landing where observations commenced in 1944.

New Meteorological Division stations were also established at Kleena Kleene and Telegraph Creek in 1942, Findlay Forks in 1945, as well as a high level station on Old Glory Mountain in 1944. Trans-Canada Airlines added new observing stations on the transcontinental airway at Penticton and Kimberley in 1941. During wartime the Grand Forks and Hudson Hope stations closed. In 1945 the Meteorological Division took over the USAAF observing programs at Dease Lake; the other American observing stations in British Columbia closed that year as did most of the RCAF stations.

YUKON TERRITORY

Synoptic observations had been taken as early as 1901 at Dawson and sent out on the Canadian Government Telegraph System. In

1939, Royal Canadian Corps of Signals operators were providing observations from Dawson and Mayo Landing. The Meteorological Division established a station in the Whitehorse settlement in September 1940 and in April 1942 the observing program moved to the airport. Watson Lake had been the only prewar observing station on the Northwest Staging Route besides Whitehorse.

As the staging route was developed the Meteorological Division opened observing stations at Snag, Aishihik, and Teslin in the summer of 1943. The only other new Canadian station in the Yukon was Frances Lake, which opened in the fall of 1941 with a Meteorological Division observer and assistance from the Hudson's Bay Company staff. Meanwhile, with their heavy air traffic, the Americans needed more observations and opened stations at Brooks Brook, Canyon Creek, Devils Pass, Fish Lake, Orchie Lake, Rancheria, Ross River, and Swift River early in 1944 and observations were commenced at Flight Strips Numbers 6 and 8 in 1945. The Meteorological Division did not continue observations at any USAAF stations in the Yukon after the Americans departed.

NORTHWEST TERRITORIES

In 1939, RCCS radio operators transmitted twice daily synoptic observations from several settlements in the NWT — Aklavik, Coppermine (Kugluktuk), Fort Good Hope, Fort Norman, Fort Resolution, Fort Simpson, and Fort Smith. On Hudson Bay and Strait observations were received from Marine Radio stations at Chesterfield, Nottingham Island, and Resolution Island. In the far North, observing stations had been established at Arctic Bay and Fort Ross in 1937.

In conjunction with the development of a Mackenzie River airway to support the Canol project, several airport observing stations were established in the territories by the USAAF. The Royal Canadian Corps of Signals stations along the Mackenzie waterway had been taking and transmitting twice daily synoptic observations since the 1920s and they continued to do so in wartime. However, as explained in Chapter 12, the Americans needed twenty-four hourly observations each day from the airports they were constructing and beginning to use. Such an observing program was commenced while the RCCS continued to provide the synoptic observations. The USAAF began sending airport reports in the spring of 1943 from Fort Resolution, Fort Simpson, Fort Smith, Fort Providence, Wrigley, Hay River, and Norman Wells. In 1944 observations were commenced at Mills Lake and Trout Lake. After the war the Meteorological Division took over the observing program at several of these airports and the RCCS stations ceased observing. During the war the Meteorological Division opened synoptic reporting stations at Fort Good Hope, Port Radium and Yellowknife.

Along the Crimson Route from Churchill to Baffin Island, the USAAF opened forecast offices at The Pas, Churchill, and Coral Harbour. New USAAF observing stations in the NWT to support flying along the Crimson Route were opened at Clyde River and Frobisher in 1942, and at Eskimo Point (Arviat), Coral Harbour, and Lake Harbour (Kimmirut) later in the war.

UPPER AIR OBSERVATIONS

For research purposes, limited amounts of upper atmosphere data had been obtained before the war from meteorographs carried aloft on kites or balloons. The instrument usually recorded temperature, pressure, and humidity. When carried by a kite the observations could be obtained when the meteorograph was brought back to earth but the height of the observation was limited. When flown on a balloon the meteorograph could travel to a greater height but when the balloon burst the instrument fell on a parachute and had to be found before the meteorological values could be determined.

Pilot balloons had been used for two decades before the war to observe the winds as the balloon ascended to the level of the cloud ceiling. The elevation and azimuth angles were periodically measured using a theodolite and by plotting these angles the speed and direction of the wind at successive levels could be obtained. Early in the 1930s, ceiling balloons were introduced. Assuming the balloon rose at a constant speed, the time elapsed before it disappeared in a cloud would indicate the height of the cloud base. Pilot balloon and ceiling balloon observations were immediately available and were useful in weather forecasting but information from meteorographs carried aloft on kites or balloons could not be obtained soon enough to be of operational value. But upper air data were required in "real time" by meteorologists to allow them to make three-dimensional analyses of the atmosphere in order to make useful aviation forecasts.

To meet this need, in the mid-1930s, aeroplane ascents were begun from twenty airports in the United States in which temperature, humidity, and pressure were measured at various levels up to about 16,000 feet (4,877 metres). (In the 1920s, Sir Frederic Stupart, then director of the Meteorological Service, had attempted to get the Canadian air force to begin such flights but he was not successful.) However, by the late 1930s, regular "apob" flights were commenced at Edmonton, Regina and Toronto and in Newfoundland. But the flights were expensive and could not take place when bad weather was occurring, the very time that the upper air data were required.

It was 1937 before the Meteorological Division's *Annual Report* mentions investigations with "radio meteorographs," the name used at first for radiosondes. Development of the concept had been rapid in the 1930s, especially in Europe and the United States, and by 1939 there was a network of several radiosonde stations in the United States. It was 1941 before a Canadian radiosonde instrument was perfected and upper air observing equipment installed at Gander, Newfoundland. More installations were made and observations commenced by the Meteorological Division in 1942 at Fort Nelson, Aklavik, Prince Albert, and Moosonee and in 1943 at Nitchequon.

In 1943 the USAAF began to establish radiosonde stations at several airfields in Canada along the eastern ferry routes and in the Northwest Staging Route/Canol sector. By January 1945 there were twenty-eight operational radiosonde stations in Canada, fifteen operated by the Meteorological Division and thirteen by the American military forces. Two ascents a day were made at most stations. Although the Canadian radiosonde was considered to be an excellent instrument it was more expensive to manufacture than the American and European models. Nor could the Canadian instruments be produced quickly enough; in 1945 half the Canadian operated stations were using instruments supplied by the Americans. Several of the USAAF operated radiosonde stations were taken over by the Meteorological Division after the war.

The chief designer of the Canadian radiosonde was Ratje C. Jacobsen who was assisted by Gerry Gill and Wendell Smith. Other pioneers in Canadian radiosonde development and operations in the war years were Larry Sobiski, Ken Hignell, and George Chapman.

RADAR STATIONS

The existence of an Aircraft Detection Corps on the East Coast in 1943 has been mentioned in Chapter 6. When this corps of volunteer observers was not activated in 1944 it was because the RCAF's chain of radar stations had been developed and eighteen such stations were activated by that summer. Most stations were along the outer coasts of Labrador, Newfoundland, and Nova Scotia. It is noteworthy that only one of these RCAF stations, Cape Bauld, in Newfoundland near Belle Isle, regularly took and transmitted weather observations.

On the Pacific coast, in 1943, there were eight detachments of a Coast Watch Unit in place along the outer coast of the Queen Charlotte Islands. This unit was disbanded by 1944 when there were eight Radar Early Warning stations in place along the outer coasts of the Queen Charlotte Islands and Vancouver Island. From records available sixty years later, it appears that none of these stations participated in the weather observing program.

CLIMATOLOGICAL STATIONS

A network of climatological stations was being organized by Professor G.T. Kingston, director of the Toronto Observatory, even before he organized the Meteorological Service in 1871. That year, he was given a grant of $5,000 which he intended to use to expand the network but the need for storm warnings became paramount and he was forced to spend most of his resources on obtaining synoptic weather observers to telegraph reports to Toronto where he developed weather forecasting methods.

But Kingston did not forget his plans to develop a climatological station network and by the end of 1871 there were more than 100 observers sending monthly reports to the new Meteorological Service. Although they were encouraged to record additional observed weather elements, temperature and precipitation were and remained the basic climatological elements reported on the monthly report forms. Observers came from all walks of life; most considered weather observing a hobby, others a service to their community, and others found the observed data to be useful in their occupations, such as farming.

Most observers were cooperative volunteers who received only the necessary instruments (thermometers, thermometer shelter, rain gauge, and snow ruler), an instruction booklet, report forms, and envelopes. In some parts of the country where no volunteers could be found, observers were paid a few dollars a month to take the observations.

In the 1870s the observing network had spread throughout the settled parts of the eastern provinces and by the early 1900s across the four western provinces. In 1901 there were 317 stations in the country and in 1931, 777 stations. That year, because of the interest in water in interior British Columbia there were 221 stations in that province, more than in any other. Because of their size and the density of their populations in the south, Ontario and Quebec ranked next with 150 and 118 stations, respectively.

During the years of World War II, the size of the national network remained fairly constant at about 1000 stations. The exact number of stations in any year is almost impossible to obtain since some observers may miss several months from sickness or travel, others have instruments for some months but fail to report, and so on. In several provinces contacts with the observers were made by provincial agents (Alberta and Saskatchewan) and in others by branch meteorological offices (British Columbia and the Maritime provinces). A provincial agency handled stations in Quebec while the observers in Manitoba and Ontario reported directly to Head Office in Toronto.

At Head Office, observations from the climatological network stations were processed and archived and some data published. Final responsibility for administering the stations in the network rested with the Climatological Section but all Head Office correspondence with observers was carried out in the name of John Patterson, director. This custom had begun in the days of Professor Kingston, continued during the directorship of Sir Frederic Stupart and existed during World War II. In the decades prior to the war the forecasters and others at Head Office had made frequent inspection trips to visit observers at both the telegraphic reporting stations and the ordinary climatological stations. With the curtailment of resources in the Depression these visits had largely ceased and, during the war, shortage of staff prevented the resumption of the inspection program.

Besides the basic observations of daily maximum and minimum temperatures and the precipitation, most synoptic and some climatological stations were equipped with such recording instruments as anemographs, bright sunshine recorders, and rain gauges. Each month records from these instruments were entered on report forms and these were also mailed to Toronto. After the war there was to be a remarkable expansion in the size of the climatological station network as the management of the stations was decentralized to the various regions in the department. By 1960 there were twice as many climatological stations in the country as there had been in 1945.

COMMUNICATIONS

In an operational weather observing and forecasting system accurate and speedy communications are essential. Weather observations have to be collected and provided to forecast offices within a time period of an hour or so and weather forecasts and other meteorological information must be delivered to the users with as little delay as possible after they are prepared. The development of Canada's storm warning and public weather forecasting program in the 1870s was only possible because telegraph services linking most cities and towns had been installed a few decades earlier. In the North, and along the coasts beyond the telegraph system, observers used wireless to send their observations to the nearest station on the telegraph system. Until the 1930s the commercial telegraph systems of southern Canada provided the observers a means for sending their observations to Toronto and Victoria. The public forecast offices in those cities used the same telegraph systems to distribute storm warnings and forecasts across the country.

In 1930, teletype was installed to link Toronto Head Office with Montreal St. Hubert airport at the time of the visit of the airship R-100 to Canada. The next year, airports in western Canada were linked by teletype to collect observations and distribute forecasts for the Western Canada Air Mail Weather Service. At about the same time a teletype was placed in the Toronto Head Office to obtain the observations received at the central telegraph office in that city and "special typists" were employed rather than telegraph operators.

With the establishment of the transcontinental airway in the 1937 to 1938 period, the Meteorological Division obtained a teletype circuit dedicated to carrying meteorological information. When war broke out this system extended from Victoria, British Columbia, to Moncton, New Brunswick, with a branch in Alberta from Lethbridge to Edmonton. The system consisted of geographical circuits connected by relay stations at Lethbridge, Winnipeg, Kapuskasing, and Toronto. Connections with the United States Weather Bureau teletype system were made at Vancouver with Seattle and at Toronto with Buffalo.

In September 1939, the system was extended from Moncton to Halifax, Montreal St. Hubert was connected to the United States Weather Bureau system at Albany, New York, and in the summer of 1940, in southern Ontario, the system was extended from Toronto through London to Windsor. Also in 1940, as noted elsewhere, the RCAF agreed to fund the connection of most BCATP flying training schools to the teletype system. As the various schools were opened and connected, the circuits became crowded and so some were divided and additional relay stations were put in place.

In the United States, meteorological traffic had increased markedly even before that country entered World War II and so, in 1941, the USWB introduced a second teletype service, labelled "Service C," to carry synoptic observations, pilot balloon and radiosonde observations, map analyses, and forecasts. It thus became necessary to connect the new Service C to the Canadian circuits at Vancouver, Toronto, Toronto Malton, and Montreal Dorval. The original system, now known as "Service A" carried the hourly weather sequences and other traffic. Over the next several months more connections were made to the American Service C at Moncton and Winnipeg and at the Dorval transatlantic forecast office.

When the United States entered the war, direct circuits were established to connect the Dorval transatlantic office with Washington and Presque Isle, Maine. Until 1942, communications between Montreal and Gander had to be by radio and the inherent interference made the service far from satisfactory. Teletype connections between Halifax and St. John's by cable became possible by July 1942 and new landlines from St. John's to Gander were soon used for meteorological traffic. With American involvement in Newfoundland additional USAAF circuits connecting the airports on the island were established. To allow the transmission of more data for transatlantic forecasting the Meteorological Division established express circuits connecting Halifax and St. John's and airports at Gander, Ottawa, Dorval, Dartmouth, and Toronto Malton. Connections with Goose Bay, however, had to be by wireless throughout the war period. When the United States decided to proceed with the Crimson or Northeast Ferry Route the USAAF installed a teletype system linking western American airports with Edmonton, Saskatoon, Regina, The Pas, and Churchill. Communications beyond Churchill were by wireless.

By 1941 the Radio Division of Transport had a wireless communications system along the Northwest Staging Route which carried meteorological traffic. With the entry of the United States into the war the USAAF erected a telegraph system along the Alcan Highway and two circuits were allotted for American and Canadian meteorological traffic. In British Columbia the RCAF put through a teletype service connecting airports on the Interior Staging Route from Vancouver through Prince George to Prince Rupert and another on Vancouver Island late in 1943. The USAAF, in 1944, established a continental service known as Service O to handle international observations and connections were made at Vancouver and Dorval with this service.

In 1944, traffic over the Canadian system out of Vancouver became so crowded an express circuit was established between that office and Calgary, Edmonton, and Lethbridge and was later extended to several British Columbia airports. Finally, when TCA decided to start a direct service from Winnipeg to Edmonton, Transport radio range stations were put in place and another meteorological circuit established.

The chief architect of the development of the Meteorological Division's communications system in wartime was Arthur J. Childs, teletype supervisor.

CHAPTER 19

HEAD OFFICE

The wartime Meteorological Division was markedly centralized. John Patterson, of course, reported to the Director of Air Services in Ottawa but all the Division's activities were administered from Toronto as they had been before the war. Andrew Thomson, the deputy controller, played a major role in the administration of the Division, especially in the aviation meteorology sector. The contributions of Reg Noble at RCAF headquarters, Pat McTaggart-Cowan at Ferry Command headquarters, Rube Hornstein and George Pincock at the RCAF forecast offices have been noted but actions they took were subject to ratification at Toronto. The senior meteorologists outside Toronto turned out to be excellent administrators and there is no evidence of significant dissension or conflict within Meteorological Division management during the war. Similarly, there are only a very few instances where Patterson was ordered by senior Department of Transport officials to alter one of his desired policies.

Fortunately, the principal activity of the Meteorological Division during the few years before the war had been the organization and staffing of a transcontinental weather service for Trans-Canada Airlines. The experience gained in this successful endeavour provided an excellent base for recruiting, training, and posting the metmen and meteorologists required by the Royal Canadian Air Force during the war. Prior to 1939 there was no military meteorological organization in place since the national meteorological service had neither responsibility nor resources for military aviation. It is significant to note that the first mention of the added responsibility occurs in the 1940–41 *Annual Report* where Patterson wrote, "The Meteorological Division is now responsible for providing a weather service for the Dominion of Canada *and for its armed forces*." [Author's italics]

ADMINISTRATION

The planning and administration of wartime meteorology in Canada was all done at the Toronto Head Office under the careful and detailed control of John Patterson. As has been noted there was little management structure in Patterson's Head Office. In retrospect he was fortunate he had a number of able and loyal meteorologists around him who accepted his leadership completely. There was no single section or unit responsible for administering programs outside Toronto but the section heads and others at Head Office administered parts of the national system in Patterson's name as well as running their own sectors in Toronto.

In the reorganization of 1936, Patterson became the Controller of the Meteorological Division and began to report directly to the Director of Air Services at the Ottawa headquarters of the Department of Transport. Administration of Division's field offices and activities across the country continued to be totally centralized at Toronto as they had been for several decades as part of the Department of Marine and Fisheries. (The department was split in 1930 and the Meteorological Service was then in the Department of Marine from that year until 1936.)

At the start of the war there were seven branch offices, four forecast centres on the airways, thirty intermediate airways stations, about 100 off-airways stations where part-time observers sent in two or three weather observations a day, 843 climatological stations, and 107 storm signal stations. The most amazing thing about this array of field offices and stations was that, in theory and in practice, there was no organization structure and each office reported directly to John Patterson, Controller of the Meteorological Division, in Toronto.

HEAD OFFICE

The paragraphs that follow describe the activities in the various areas of responsibility at Head Office.

RESEARCH

Both the Meteorological Division and its predecessor, the Meteorological Service of Canada, had been established to provide storm warnings and weather forecasts as well as climate information services and so had few resources for meteorological research. What research that took place was done in the Physics Division whose head was John Patterson. He was keen on instrument research and had developed a new three-cup anemometer and a portable barometer for use in the Service. Andrew Thomson succeeded Patterson as the atmospheric physicist and was more interested in climate change, sunspots, and lunar atmospheric tides than instruments. Once the war began both Patterson and Thomson were too engrossed in administration to participate in any research.

In 1934 Graham Millar, a graduate of the first MA course, was hired as a research meteorologist to undertake evaporation and lake water studies and was put in charge of marine observations. In 1936 W.E.K. Middleton and Millar published a paper on the temperature profiles across Toronto using data obtained from mounting a fast-response electric thermometer on Millar's car and driving twelve kilometres across the city several times in summer and winter. The paper is considered a classic example of early research in the study of city climates.[90]

Just before the outbreak of war, Millar's Great Lakes studies had to be put aside when he was given the job of developing and supervising the necessary code and cipher work at Toronto. After the war, Canada's seagoing ships moved to foreign flags and Millar was free to return to his Great Lakes research. In 1952, he published an extensive paper on the surface temperatures of the Great Lakes before he resigned from the Division to go to the Naval Research Establishment in Halifax.[91]

Dr. Bernhard Haurwitz, a noted European physicist and a refugee from Nazi Germany, became a Carnegie Fellow at the University of Toronto in 1935. While waiting for the Meteorological

Division to obtain permission to hire him as a non-national, he worked with the Division, gave the dynamic meteorology lectures to successive MA courses, and acted as an advisor to students in their research projects. During this time he published his lectures on the physical state of the upper atmosphere in book form before leaving for the United States in 1941.

Wendell Hewson, a Toronto MA graduate in meteorology who had obtained a fellowship to the University of London, returned to the Division in 1938 with the first Canadian Ph.D. in meteorology. That year he was delegated, with Gerry Gill, to undertake field research in the Columbia River Valley of British Columbia where sulphur fumes emanating from ore smelting at Trail were damaging crops to the south in the United States. Their report to the joint Canadian and American Trail Smelter Arbitral Tribunal, the first major Canadian study of atmospheric pollution, was published in 1944.[92] A further contribution by Hewson was a textbook, *Meteorology Theoretical and Applied*, published with Dick Longley and used in wartime and postwar training courses.[93]

Although the wartime meteorologists had little time for experimental nomograms or research, several forecasters devised and prepared forecasting aids in the form of tables using climate data and weather maps. Others, especially those engaged in training at Head Office, continued to seek better methods of analysis and forecasting. Bill Thorn undertook extensive studies in isentropic analysis although this was not often used in the forecast offices. Clarence Penner, Al Crocker, and Warren Godson began their research in synoptic map analysis, which led to the publication of several papers on the Canadian three-front model and frontal contour analysis after the war.

INSTRUMENTS

As mentioned above, John Patterson, wartime Controller of the Meteorological Division, was an excellent instrument man as well as an atmospheric physicist. In the 1910 to 1929 period, he had developed a meteorograph, an instrument with sensors carried aloft on a kite or a balloon to measure and record temperature and pressure, a transportable barometer that became known as the Kew-Patterson barometer, and a three-cup anemometer which became the Division's standard for some time. Late in the Great War, Patterson had been seconded to the British Admiralty Board, with University of Toronto professors, to work on methods of extracting helium from natural gas.

Middleton, hired in 1930, followed Patterson as the Division's instrument specialist. He gave the instrument courses to the MA and subsequent wartime courses for meteorologists and metmen. A major contribution to international meteorology was his textbook *Meteorological Instruments*, the first general textbook in the English language on the subject in more than fifty years.[94] During the war he undertook development work on an early automatic weather observing station which was stationed in Lake Ontario for two summers. The experimental readings were transmitted by radio for interpretation at a nearby RCAF air training station.

The Meteorological Service had participated in the 1932–33 International Polar Year by establishing and manning three temporary observatories in the Arctic. Experimental radiosonde observations were taken at one of these stations (Coppermine) and from that time through the war period, Patterson at Toronto gave high priority to radiosonde instrument development and calibration work. R.C. Jacobsen was the leader in this applied research and the Division manufactured and calibrated the instruments.

CLIMATOLOGY

Abraham J. Connor, appointed Dominion Climatologist in 1911, was still the section head in wartime. Because of Connor's ill health a few years earlier, Clarence Boughner, a 1934 graduate of the first MA course, was brought into the section and by the 1940s was effectively in charge of its work. Because of the need for space for the wartime training program in the Head Office building (315 Bloor Street West), the Climatological Section was moved in 1941 to the nearby old McMaster College building. Within a year or so the section moved again to the Medical Arts Building, also very close to the Head Office. Mr. Connor refused to leave the 315 Bloor Street building and continued to work from an office that he shared with the librarian.

During the war years the Climatological Section continued to collect, on a monthly basis, the synoptic and climatological observations taken throughout the country. The observers in Manitoba and Ontario sent their observation forms directly to Toronto each month while observers in the other provinces sent theirs to a branch office or a provincial agent where initial quality control measures were taken before the forms were sent to Toronto. (The Quebec Streams Commission handled this task in that province.)

In the Climatological Section the observations were processed, given final quality control measures, and archived. Staff shortages caused the regular data periodicals publishing program to fall into arrears by several years by the end of the war. The branch offices maintained contact with the cooperative climatological observers, supplied them with observing forms, prestamped addressed envelopes, and replacement instruments but the opening of new stations and the selection of new observers was administered from Toronto.

The *Monthly Weather Map*, a publication started in 1895, was published each month with a delay of about five weeks for the collection of the report forms and summaries from the branch offices. Besides national maps of temperature and precipitation each issue contained tables of mean and extreme temperatures and precipitation from most climatological and synoptic stations. By the fall of 1945 the *Monthly Record of Meteorological Observations in Canada* (a publication commenced in 1916), which contained quality controlled daily and monthly data, had slipped into arrears by five years. During the growing season a *Weekly Weather Summary for the Prairie Provinces* was issued in co-operation with the Winnipeg Grain Exchange weather office and a *Monthly Summary of Degree-Days* was published in season for the heating and energy sectors.

The only publication added during wartime was the *General Summaries of Hourly Weather Observations in Canada*, first compiled and published in 1944 at the request of aviation interests. It contained monthly tables of weather types, cloud amounts and types, obstructions to vision, etc., and tables showing cloud ceilings, visibility, and wind directions by speed groupings prepared clerically for a selection of airports in Canada. Also published in wartime were a few studies of regional climates that had mostly been prepared before 1940.

Mr. Connor and his secretary handled requests for climate data and information from the public, industry, and commerce throughout the country. In the early war years there were many requests for wind data from engineers in the Transport and National Defence departments who were engaged in planning runway orientation and construction at the many new airports being built for the Royal Canadian Air Force. Later in the war United States military and civil offices made frequent requests for climate information to assist in planning the many facilities they constructed in northern Canada. The United States Weather Bureau began a program of microfilming observation report forms from northern Canadian stations which was continued and extended by the Meteorological Division after the war.

The office staff, supervised by Boughner, became understaffed and numbered less than a dozen by 1945. A few men had departed for war work and the remaining staff was mostly clerical and female. A veteran senior technician, Peter Kerr, was responsible for some aspects of clerical data processing and he was joined during the war by James Kirk, who had been stationed at St. John's to supervise observations at the Newfoundland stations before the war. Two or three typists were kept busy typing data for the publications while clerks logged in the observation report forms, used adding machines to complete and check the summaries, filed the forms, proofread data manuscripts and so on.

WARTIME CIPHERING

On account of the growing war threat, John Patterson was ordered by his department in January 1939 to have cipher books prepared for use at those weather observing stations where observations were transmitted by wireless. Graham Millar, a research meteorologist, was asked to prepare cipher books to be sent to the Arctic stations with their annual mail at the time of the summer supply missions. This was done and so when war was declared the stations were ordered by wireless to begin using the codes.

At that time, Millar was made responsible for all coding and cipher work at the Toronto Head Office. Besides handling the deciphering of the observations, all reports and forecasts transmitted from Toronto for the navy and other users had to be enciphered. He was authorized to build up a sizeable staff of men and later women who worked in shifts around the clock. Millar, who was particularly precise and thorough, became an expert on ciphers and invented simple formats for enciphering forecasts for ships on the Great Lakes. He also handled the Head Office correspondence having to do with compliance of radio and the press in suppressing radio weather reports and forecasts. (Millar's account of his years in the Meteorological Service, cited above, has been valuable in assisting the author to portray activities at Head Office in wartime.)

PUBLIC WEATHER FORECASTING

The sixty-year-old public weather forecasting system was left unchanged during World War II. The forecasting program had been initiated during the 1870s by Professor Kingston, director of the Toronto Observatory and superintendent of the new Meteorological Service of Canada; storm warnings and public weather forecasts were prepared and issued from the Map Room as early as 1876. Since the observing network extended no further west than Fort Garry (Winnipeg), the first forecasts were limited in geographical extent to eastern Canada. With the settlement of western Canada and the extension of telegraph facilities to the Pacific coast, a smaller forecast office was established at Victoria, British Columbia's Gonzales Heights Observatory, in 1898. By 1903 weather forecasts were available every day in all parts of southern Canada.[95]

There is no record of what improvements John Patterson had planned for his public weather forecasting system once he had established the required aviation forecasting program. It is significant, however, that in the two years or so before the war, he did not use any of his newly trained forecasters for public weather work. These forecasters, only one with a college degree, were not trained in scientific air mass analysis but used empirical methods that had been developed decades ago in Washington and Toronto. Experience was important; it was considered that at least three years apprentice experience was necessary before one could prepare and issue a forecast.

Assistants plotted observed station data on the weather maps and then the forecaster inspected the map and drew isobars. He marked the areas of high and low pressure and considered how they might develop and move over the next twenty-four hours or so. Usually a second map was plotted with temperature data and isotherms were drawn. No use was made of the new radiosonde data and the only forecasting tools used were the files of historical weather maps and the current temperature map. After current use the weather maps were bound and filed chronologically. In considering the current situation, the forecaster would often look for a similar situation in the past, see what happened in the next day or so, and use this as guidance in his forecasts of tomorrow's weather.

The wartime public forecasters at Toronto were Frank O'Donnell (chief), Karl Fox, Melvin Bessie, and Ernest Chilcott. They prepared

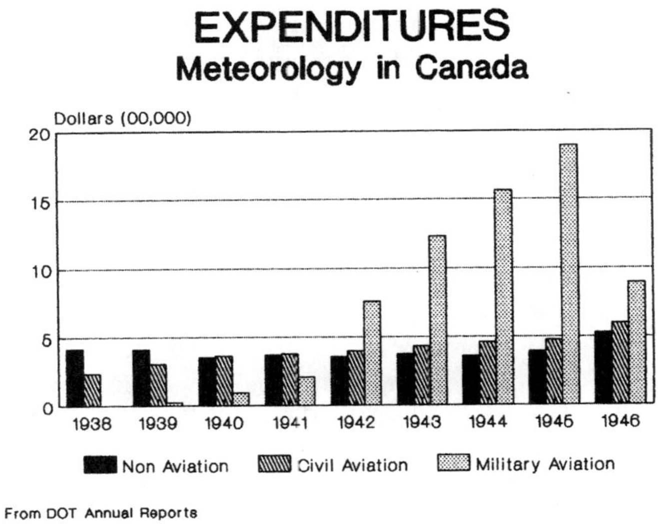

FIGURE 10

Annual federal expenditures in Canada for civil aviation meteorology, for military aviation meteorology, and for non-aviation meteorology. Non-aviation meteorology includes Head Office administration and other activities in Toronto, stations in the civil observing network, communications, and public weather forecasting.

forecasts for several regions across southern Canada from the Rockies to the Maritime provinces and Newfoundland. Forecasts were issued twice daily, in the morning and evening, for twenty-four hours in each case. At the Victoria Gonzales Heights Observatory Bill Thorn had been in charge of forecasting for southern British Columbia and his assistant William Burton carried on for a while when Thorn returned to Toronto.

Shortly after the Newfoundland transatlantic forecast office was established before the war that office began to prepare and issue the Newfoundland public weather forecasts, storm warnings for the island, and special forecasts for some resource industries. The task of preparing special forecasts for fruit farmers during the spring frost and spraying seasons in the Annapolis Valley was given to the new Halifax forecast office in 1942. That year, on the West coast, public forecasting responsibilities were given to the new Western Air Command forecast office. As the war progressed and weather security measures were relaxed, a few other aviation forecast offices began to provide local newspapers and radio stations with forecasts but the forecasters in the Toronto Map Room continued to be responsible for all public forecasts in central and eastern Canada.

It was after the war, in the spring of 1946, before the public weather forecast office was disbanded in Toronto. O'Donnell retired and the other forecasters and assistants were moved to jobs at Toronto Head Office and the Malton forecast office. Both the Eastern and Western Air Command forecast offices were converted to Dominion Public Weather Offices (DPWOs) while the District Aviation Forecast Offices at Edmonton, Winnipeg, Toronto Malton, and Montreal Dorval became combined DPWOs and DAFOs.

TRAINING

Patterson and his assistant controller, Andrew Thomson, had been responsible for the prewar MA course training with help from Haurwitz and Middleton and from University of Toronto professors. Des Kennedy was brought back from Winnipeg in September 1939 to assist and when the wartime courses began Jim Leaver returned from Trenton/Rivers in December 1940 to help. These two meteorologists were directly responsible for training in the short and advanced courses over the next several years.

Middleton continued to give the instrument course and others who participated in the wartime training at Head Office were Wendell Hewson who lectured in dynamic meteorology, and Richmond Longley who joined the training group from Halifax in September 1941 to teach synoptic meteorology. Al Crocker was brought back to Toronto to instruct in synoptic map analysis in August 1942. The only lecturer from the university in wartime was Professor Griffith Taylor who lectured in climatology. To assist in supervising the map work laboratories a meteorologist was frequently posted to Head Office for a few months and, on other occasions, a meteorologist was given such duties for several weeks after illness to avoid shift work at an operating forecast office.

The meteorologists listed above also played a major role in selecting metmen and meteorologists for posting to the various stations. They also wrote the sometimes necessary critical letters to metmen in the field for signature by John Patterson. No one knew the trainees better than their instructors and they were the ones who conferred and decided on posting lists after each course and probably on the movement of metmen from station to station. New metmen who were unhappy with their postings were sent to see Andrew Thomson. He usually allowed some changes to be made and kept the metmen reasonably happy. Also, the instructors probably constituted unofficial rating boards to select metmen for the advanced courses from those who had applied. After 1941 an experienced metman, Fred Patterson (no relation to John Patterson), looked after the posting details as metmen were moved from station to station to meet the varying needs of the day.

OTHER ACTIVITIES

Alvin Thiessen became the Division's librarian at about the time the war began. He was the office's second professional librarian. His predecessor, Miss H.J. Hotson, had been hired in 1936 but departed after two years or so. Thiessen left the Division towards the end of the war to organize a library for the Provisional International Civil Aviation

Organization (later ICAO) that was being organized in Montreal. John Patterson's secretary during the war years was Leroy Jones and then Evelyn Self. Berneice Hall was Andrew Thomson's secretary.

Stanley Bane and Mervyn Watson were the instrument makers and Alex Burnside a helper. Arthur "Major" Condie was the office printer and "Duke" Hennessey was the map draftsman. Fred White was the carpenter and E.J. Franklin the accountant. The teletype system was headed by Art Childs and the weather observations at Toronto were the responsibility of Murray Monsinger and James McPherson.

After he returned from Winnipeg in 1943, Don Archibald administered airway observing. Frank O'Donnell supervised the network of storm warning stations and the traditional synoptic observing station network. (The forecasters had done most of the station inspection before the war.) Connor and Boughner looked after stations in the national climatological observing station network.

Except for Don Archibald who was the Western airways superintendent until 1943, the officers-in-charge of the District Aviation Forecast Offices had no formal responsibilities beyond their own offices. However, in practice, they were called upon from time to time to help solve regional and local problems for the Division. Likewise, the OICs of the two RCAF forecast offices (Hornstein and Pincock) had no formal responsibilities for other meteorological sections but the RCAF Commands increasingly looked to them for advice and to solve their meteorological section problems. The same situation existed in dealing with transatlantic aviation meteorology where Pat McTaggart-Cowan was both Patterson's officer-in-charge and Principal Meteorological Advisor to the head of the Royal Air Force Ferry Command.

[90] W.E.K. Middleton and F.G. Millar, "Temperature Profiles in Toronto," *Journal of the Royal Astronomical Society of Canada*, vol. 30, no. 7, pp. 265–272.
[91] F. Graham Millar, *My Years in the Meteorological Service*, Halifax, August 1992.
[92] E.W. Hewson and G.C. Gill, *Meteorological Investigations in the Columbia River Valley, near Trail, B.C.* (Report submitted to the Trail Smelter Arbitral Tribunal) Bulletin No. 453, United States Department of Interior, 1944. pp. 23–228.
[93] E.W. Hewson and R.W. Longley, *Meteorology Theoretical and Applied* (New York: McGraw Hill, 1944).
[94] W.E.K. Middleton, *Meteorological Instruments* (Toronto: University of Toronto Press, 1941).
[95] See Morley Thomas, *Forecasts for Flying*, pp. 38–43.

CHAPTER 20

PLANNING FOR PEACE

In March 1944, when the British Commonwealth Air Training Plan agreement expired, there was a surplus of aircrew, flying training was reduced, and some RCAF air training stations were closed by the end of that year. There was not as yet a surplus of metmen and meteorologists but it did become possible to bring the meteorological strength up to establishment at the continuing schools and to have metmen available for tasks that had begged for attention for years.

Although much planning for peace must have taken place in 1944 and 1945 at the Meteorological Division Head Office in Toronto only a few documents can be found on the subject today. However, some information about an Interdepartmental Committee on Meteorology and a 1945 Conference of Meteorologists, both having to do with planning for peacetime, do exist. And, since a new Head Office organization was announced in 1946, much discussion and planning must have taken place in Toronto during wartime.

METMEN AND METEOROLOGISTS

Individual meteorologists and metmen certainly began to plan for peace as early as 1944. Only a few had been hired before the war and had obtained permanent status in the civil service. Everyone hired during the war was employed on a temporary basis and many began making plans to return to the professions that they had left when joining the Meteorological Division. Some had been schoolteachers or university lecturers and, in most cases, their jobs were waiting for them. A few were released from the Meteorological Division in time to begin the school year in September 1944 while most former teachers left in time to resume their teaching careers in September 1945.

For those meteorologists and metmen who had joined directly from universities, it meant considering the possibility of applying for graduate school admission or of finding a position in a discipline close to their prewar education and training. And, a surprising number of meteorologists and metmen decided to remain in meteorology if at all possible. All were aware that the Division had expanded tenfold in wartime and all had been warned that there was no promise of permanent positions for any of the wartime recruits once the war was over.

A majority of those who had taken both the intensive and advanced courses elected to remain with the Division; of the 125 or so meteorologists who were trained in wartime, 100 remained in meteorology until the early 1950s and ninety or so remained for their entire careers. This was probably not too unusual since of the fifty MA meteorologists hired before or early in the war, only five departed at the end. No records are available showing to what professions the departing meteorologists went but at least half a dozen went to universities to teach and become professors, others went into law, some returned to school teaching, and others to private business. The metmen, not as well qualified for a professional career in meteorology as the meteorologists, left the Division in great numbers in late 1944 and 1945; only 110 remained in 1946 of the 220 who were at RCAF stations in March 1944.

COMMITTEE ON METEOROLOGY

In 1944, the War Cabinet approved, by order-in-council, the formation of a permanent Interdepartmental Committee on Meteorology. The reasons advanced for such a committee were based on the great diversity of interests served by the Meteorological Division and the ramifications of meteorological programs on industry and government. It was pointed out that meteorology had become increasingly important to many fields of activity in the country — agriculture, aviation, the Armed Services, construction, forestry, hydroelectric power, to name a few, as well as to the general public.

The committee reported to the Minister of Munitions and Supply, then C.D. Howe, and the secretary was a member of Privy Council Office. The membership consisted of the Deputy Minister of Transport, the Controller of the Meteorological Division, representatives from the Air Services and Marine Services branches of Transport, and members appointed by Ministers of the other concerned departments — Agriculture, External Affairs, National Defence, Mines and Resources, and the National Research Council.

The committee was made responsible for acting as a medium for consultation between departments and agencies on questions regarding meteorological services. It was given authority to recommend on matters relating to meteorological services required by other departments, on meteorological subjects requiring investigation and research, and to make recommendations on appropriate forms of cooperation with international organizations. It should be noted that at this period, late in the war, meteorology and meteorological services were held in relatively high regard in civil government and military agencies.

There had been favourable publicity given to meteorology in wartime. It was associated with the success of the British Commonwealth Air Training Plan where tens of thousands of aircrew had been given training in meteorology and who were well aware of the effect weather had on their training and operations programs. Also, for security purposes, weather forecasts had not been made available to ordinary Canadians for some periods during the war and there were expectations of improved forecasts that meteorology would provide to civilian society after the war. What was not recognized was that in the war years great sums of money were spent on meteorology, money that would not be available in peacetime to give the public a weather service equivalent to that which RCAF had received. Another factor that mitigated against the success of the committee was that the senior meteorological officials could not have been anxious to have officials from other departments advising how the Meteorological Division should spend its budget.

Finally, the possibility that the committee would provide advice on appropriate forms of cooperation and participation in international meteorological affairs was probably not too well received by the Division officials. At the end of the war, international meteorology was about to be thoroughly revamped by converting the venerable International Meteorological Organization, a club of directors, into the World Meteorological Organization where the heads of national meteorological services spoke for their countries. In fact, the Convention of the WMO, which came into effect in 1952, named the national head of meteorology to be the Permanent Representative of his country without reference to any external or foreign affairs agencies.

While the 1945–1946 *Annual Report* of the Division carried a few paragraphs on the new Interdepartmental Committee, no mention was made of it in subsequent reports. After a very few meetings the committee became dormant and did not play a significant role in planning for Canadian meteorology in peacetime. The Meteorological Division did not, however, turn its back on the requirements of the other departments. Beginning about 1950 the Division began to second meteorologists to other government departments to work on their meteorological problems, with the Division continuing to pay salaries and the host departments paying the travel and other costs. The secondment policy was to exist for about two decades; about a dozen positions were established to allow professionals to be seconded to Agriculture, Forestry, National Research Council, British Columbia Forestry, Ontario Conservation, and the Prairie Farm Rehabilitation Administration.

CONFERENCE OF METEOROLOGISTS

In 1945, the Department of Transport, perhaps urged by Trans-Canada Airlines, authorized a study of means to improve the peacetime airways meteorological service. Mr. R.T. Moore of the Department of Transport and Mr. D.C. Archibald of the Meteorological Division were given authority to conduct a survey of the

Meteorological Division and to make recommendations. To assist them, the officers-in-charge of the main forecast offices — Vancouver, Edmonton, Lethbridge, Winnipeg, Toronto Malton, Montreal Dorval, and Moncton — were circularized early in June 1945 asking for their views on the subject with recommendations for improvement to services for commercial air transport.

The meteorologists were asked to make their contributions under such subject headings as office accommodation, meteorological information (observations), monitoring (constant watch of the hourly and synoptic observations on the teletype), standardization and development of methods of analysis and methods of presentation of meteorological information, methods of improving the accuracy of forecasts, educational aids for meteorological personnel and for flight personnel, and so on.

Lengthy recommendations and comments from those invited were tabled at a Conference of Meteorologists held in Winnipeg, June 15–19, 1945. The responses from the meteorologists varied considerably both in length and substance and consisted of commonplace complaints over lack of space and proper furniture, poor lighting, insufficient time allowed for map analysis and forecast preparation, lack of observations, lack of liaison with TCA, lack of shift overlaps, and such which might be expected after several years of wartime shortages in personnel and equipment. One began with the comment that the request for suggestions for improvement implied criticism of the present service. He explained that in 1941 meteorologists were told that individual trip forecasts were unnecessary except in special cases but the pilots, when given only regional forecasts, often complained. The meteorologist reported that the pilots had the sound belief that regulations still called for trip forecasts and that the meteorologists were neglecting their work when none were prepared and given to them.

Some meteorologists stressed the need for standardizing methods of analysis and forecast presentation but others pointed out that methods could not be rigorously standardized without loss of efficiency. For example, each meteorologist tended to develop his own methods and must be allowed to employ them when preparing forecasts. Many suggestions were given regarding education and training:

for the meteorologists, better DAFO libraries, regular local seminars, visits to adjacent forecast offices, refresher courses at Head Office, visiting tutors in meteorology, and more familiarization flights on TCA scheduled flights; for the flight personnel, a correspondence course and lectures by the meteorologists. Nearly all respondents stressed the need for more staff and some suggested salary ranges and space allocations for the different classes of professionals and assistants.

It is interesting that only one response went beyond the terms of reference for the conference and suggested a new organization plan not only for a region or district but also for the Meteorological Division as a whole. It is apparent that this widening of the area for consideration met with favour at the conference since the report of the conference begins with a paragraph that reads "The providing of adequate meteorological information for any particular activity is dependent on the complete meteorological organization . . . it is contended that the authority . . . to conduct a survey of the Meteorological Service should not be restricted to those portions of the organization which are directly concerned with Air Transport operations." Following this justification the report contained a recommended outline of a national organization from the executive headquarters through regional offices, central forecast offices, and dependent offices. It went on to justify and outline the responsibilities and functions of the different levels of the national organization in forecasting, observing, education and research, climatology, personnel and communications, with suggestions for staffing.

There was one minority report filed with the conference report. That was from A.J. Childs, teletype supervisor, who objected to the recommendation that a regional teletype supervisor should report to a regional officer rather than to the national teletype operations superintendent at Head Office. D.M. Robertson, officer-in-charge of the Winnipeg District Aviation Forecast Office and chairman of the conference, wrote to Messrs. Moore and Archibald on June 18, 1945, enclosing copies of the "Report of Recommendations of Meteorologists." Two weeks or so later N.J. Humphrys prepared a "Meteorological Report by Trans-Canada Airlines," which was probably also sent to Moore and Archibald. Both documents are remarkably similar in content; the TCA document contained excerpts from the conference document with added justifications for many recommendations. Neither report was ever published.

It is evident, half a century later, that the meteorologists were used by some, certainly some in Trans-Canada Airlines, who wished to see the Meteorological Division taken over and converted into a totally aviation-based organization, probably at the expense of the two senior officers of the Division, John Patterson and Andrew Thomson. But Patterson, then sixty-seven years of age and about to retire in a year or so, and Thomson, then fifty-two and the heir apparent to the directorship of the Meteorological Division, had no intention of allowing the Division to become massively aviation oriented. They doubtlessly found support at senior levels of Transport and other departments interested in meteorology because the sweeping recommendations from the conference were never adopted and, in fact, no record of any documentation or recommendations can be found in the Division archives.[96]

It is interesting to see that some of the improvements suggested by the meteorologists did take place in the immediate post-war years. In October and November 1945 a six-week refresher course was held at Head Office to which the officers-in-charge of the various forecast offices were invited along with a few other selected senior meteorologists. While the course was primarily to bring the meteorologists up to date on synoptic weather analysis and forecasting it also served as a forum for discussions and planning to improve services. Included in the course was a set of lectures on personnel management given by a university professor. And, of course, better accommodation and larger staffs did come with peacetime as budgets grew and Transport built new airport terminal buildings in which the forecast offices were usually accommodated.

USAAF STATIONS

As noted in earlier chapters, the United States Army Air Force was given permission, under the North American defence agreement, to establish observing and forecasting offices in Canada when needed. Several score offices were built and staffed by the USAAF from 1942 to 1944 and by 1945 many were no longer needed. As the Americans left, the buildings and equipment reverted to the Canadian War Assets Corporation for disposal. The Meteorological Division held meetings with the USAAF at which the possibility of Canada taking over some offices was discussed and a senior meteorologist (Des Kennedy) was sent to inspect the offices along the Crimson Route to report on their equipment and state of repair.

The considerations regarding what observing stations to take over probably involved the most difficult planning exercises for John Patterson and his staff. Many American stations were in vital areas where observations were greatly needed but there was no possibility that the peacetime Division budget would expand to the degree necessary to operate them all. The result of this planning was that most American stations were allowed to close when the Americans left but ones at airports along the Mackenzie River, on Hudson Bay and Baffin Island were retained, some to be manned by Meteorological Division observers and others by Marine and Radio Range station operators.

PATTERSON'S RECOMMENDATIONS

After he retired in 1946 John Patterson undertook to record the history of the Meteorological Division in wartime. He had a desk in the "shop," the addition to the original instruments shop, which later became the library. Patterson completed his manuscript in the early 1950s but it was deemed unpublishable by reviewers and the administration agreed. Fortunately the manuscript was saved and it has been of great value in writing this book.

Patterson's final chapter had to do with recommendations of what actions should be taken by the Meteorological Division when the next war occurred. By the time Patterson wrote this chapter confrontations with the USSR were occurring and he believed there was a very good chance that Canada would be one of the war theatres if the cold war became hot. The recommendations summarized below reflect Patterson's opinions formulated from his experiences during World War II.

a) Women had proved to be competent and should be employed where ever possible and given the same opportunity for training as men.

b) The Meteorological Division should be militarized especially if Canada became a theatre of war. (In December 1945 the Deputy Minister for Air of the Department of National Defence advocated that a number of metmen be commissioned in the RCAF and placed on reserve to be called up when needed for periods of two years.)
c) The Division should ask for authority to hire and train metmen beyond immediate needs so as to be ready for emergencies.
d) Training courses should be given in summer for meteorologists in reserve status.

(The Meteorological Division continued to provide meteorological services to the RCAF in peacetime and only those who served overseas were commissioned.)

METEOROLOGICAL DIVISION REORGANIZATION

Planning for the reorganization of the Meteorological Division began during the war but could not be completed until fiscal year 1946–1947 due to retirements and the transfer of some senior staff to and/or from Head Office. When the reorganization did occur all meteorological work was organized into six sections — Forecast, Planning and Administration, Training and Research, Instrument, Climatology, and Basic Weather with the heads, then called assistant controllers, located at Head Office in Toronto.

In addition, special liaison offices were established with Trans-Canada Airlines at Winnipeg and with Air Services Branch at Ottawa. This organization was largely maintained for decades indicating that the reorganization planning by Patterson, Thomson, Noble, and McTaggart-Cowan was sound. The name of the Division was changed to the Meteorological Branch and then briefly to the Canadian Weather Service before it was moved from the Department of Transport to the new Department of the Environment and given a new name, the Atmospheric Environment Service in 1971. Through these moves little change was made in the internal organization until major changes were made to conform to new government policy in the

1980s and 1990s. (In late 1999 the name Atmospheric Environment Service was dropped and the service became known as the Meteorological Service of Canada.)

[96] AES, Archives, History of Canadian Meteorology File, "Report of the Recommendations of Meteorologists" and "Meteorological Report by Trans-Canada Airlines," 1945.

APPENDIX 1

315 Bloor St. West,
Toronto 5 . Ontario
September 26, 1941

Dear xxxx;

[I have been informed] that you are considering applying for a position in connection with the activities of the British Commonwealth Air Training Plan for the duration of the war.

It would appear from your University training in Mathematics and Physics that you are especially qualified for this branch of work. Those who are appointed as Meteorological Officers serve in a civilian capacity at air training stations. Prior to being assigned to an R.C.A.F. school, successful applicants are given an intensive three months' course in meteorology at Toronto, after which they are assigned for a short period to an airways District Forecast Office such as Malton or Winnipeg.

The duties of a Meteorological Officer are to give lectures in elementary meteorology to the student pilots and to prepare weather forecasts for the operations personnel, based largely on the weather advice sent from the larger airways forecast centres.

The base salary paid is $135.00 a month for the duration of the war and begins from the time the successful applicant commences his course at Toronto. A cost of living bonus and an additional $30.00 a month grant is made during the time a meteorological officer is stationed at an air training school.

J. Patterson

APPENDIX 2

THE WARTIME METEOROLOGICAL COURSES

a) The 15-week Intensive Course in Meteorology
(short course)

Elementary meteorology (given in the first few weeks)	45 hours
Synoptic meteorology	42
Dynamic meteorology	42
Climatology	11
Meteorological instruments	14
Physics of the Upper Atmosphere	6
Aerial navigation	4
Weather map analysis and forecasting (each week: plotting 5 hours, analysis and forecasting 10 hours, discussion 5 hours)	300
Practice lectures	20
Practice observing	6
Teletype procedures	3
Current map discussions	50

b) The 15-week Advanced Course in Meteorology

Dynamic meteorology	45 hours
Thermodynamics and radiation	45
Synoptic meteorology	30
Meteorological instruments	30
Climatology	15
Upper air	60
Map analysis and forecasting	360

APPENDIX 3

WARTIME METMEN AND METEOROLOGISTS

TABLE 1

Graduates of the Wartime Short Intensive Courses in Meteorology

+ Those who subsequently completed the Advanced Course

* Those who continued in meteorology for at least several years after the war.

No. 1 Course November 1940–February 1941

G.B. Brown+*, W.F.M. Buscombe+, D.B. Currie+*, J.D. Dobson+, W.R. Fryers*, E.A.S. Galbraith*, L. Hillgartner+*, A.A. Hoover+*, K.M. Korven*, J. Lauder*, E. Longhurst, A.B. Lowe+*, A.G. MacVicar+*, D.N. McMullen+*, B. McElheran, J. Moncur, R.F. Pattison*, P. Perry, H.V. Tucker+*.

No. 2 Course March 1941–June 1941

G.P. Adamson, J.E. Blanchard, E.A. Brickman*, G.M. Busche*, L.A. Cooke+, R.V. Dexter+*, M.R. Fleming+*, R.W. Glen+, F. Goodwin, W.L. Gutierrez+*, A.W. Jackson+*, N.V. Jefferson+*, E.A. Johnston+*, A.H. Mason+*, J.H. Meek, D.E. Page+*, W.K. Sly+*, W.C. Thurber*, W.J. Wallace, E.D.M. Williams*, C.H. Sutherland+*, A.L. Wright.

No. 3 Course June 1941–September 1941

H.W. Baynton+*, K. Buckthought+, R.D. Butterill+, D.G. Chapman+, G.M. Chapman*, S.W. Dewar+*, R.H. Douglas+*, J.A. Ferris, W.W. Fraser, H.M. Greb+, E.J.A. Hamilton*, J.D. Holland+*, H.M. Hutchon+*, P. Johns+*, J.L. Knox+*, G.H. Legg+*, A.P. Loomis, R.C. McKenzie, A.F. McQuarrie*, R.B. Munro, G.T. Munn, R.E. Munn+*, J.D. O'Reilly, J.G. Potter+*, R.D. Ross, W.W. Stewart*, C.M. Thompson, A.F. Vogt.

No. 4 Course October 1941–January 1942

E.A. Barks+*, G.C. Cameron+, W.M. Cameron+, B.S.V. Cudbird*, D.C. Day+*, W. Halina*, W.R. Jarmain, E.H. Jones, J.E. Little, A.C. Longman, L.K. McGlening*, W.E. Moore, F.A. Page*, T.L. Richards+*, D.F.A. Slater+*, D.H.M. Stewart, M.K. Thomas*, R.L. Titus+*, H.S. Vaughan, R.W. Walkden+*, B.G. Walliser, D.J. Wright*.

No. 5 Course February 1942–May 1942

W.W. Allen+*, G.F. Allison+, B.W. Boville+*, P.R. Brun+*, W.G. Clark+, W.R. Cunnington, A.J. Ducklow, G.R. Kendall+*, N.A. McDonald, D.G. McCormick*, D.A. McLarty, J.R. Miller+*, L.H. Slind, R.O. Spence, L.G. Tibbles+*, xxWesley, J.P. Wilkinson, A.B. Wright+, J.B. Wright+*.

No. 6 Course June 1942–August 1942

T.V. Adams, K.E. Andrews, F.N. Beard, H.C. Belhouse+*, C.G. Black, D.G. Black+, J. Bocking, J.A. Burgess+*, J. Calder, L.T. Campbell+*, H.G. Capelle+*, G.W. Clarke, G.L. Clendenning+, R.H.D. Conn, W.E.H. Cross, G.E. Cutler, D.M. Davies, E. Delisle, E.H.V. Dexter+*, R.R. Dodds+*, E. Einarsson+*, E.N. Ellis+*, E.M. Elsley+*, E.M. Fleming+*, L.B. Foster*, T. Fostvedt, W.F. Ganong+*, W.L. Godson+*, G.H. Hall, W.R. Hamilton+*, K.F. Harry+*, J.K. Heise, J.R. Henderson+*, K.A. Hignell*, W. Hoddinott+*, J.R. Hood, A.F. Ingall+*, M. Jenkins, H.S. Keenleyside+, E.J. Kermode*, R. King, A.H. Lamont*, T.H. Lopson, S.B. MacKay, W.H. Mackie*, A.R. McCracken+, R.V. Maclean, F.R. Mahaffy*, W.E. Markham+*, D.A. Moddle, L.R. Mumford+, G.H. Muttit+*, T.A. Nixon, B. Parkhouse, F.M. Riddle, R.J. Rutherford*, F.G. Scott, V.E. Solman, W.M. Sprules, F.L.V. Taylor, R.V. Tyner+*, O.K. Van Sickle, W.S.C. Wallace*, J.R. Wilson, G. Wilson, N.D. Williams.

No. 7 Course September 1942–December 1942

A.E. Allison+*, W.W. Bates, J.C. Beatty, C.J. Bishop, J.M. Blackhall, C.F. Break, N.R. Brining, W.J. Callum, R.H. Craddock+, R.H. David, H.G. Dearborn*, T.A. Elliott, W.K. Etherington, T.E. Gutier, C.G. Goodbrand+*, S.V.A. Gordon+*, L.G.F. Gray,

T.A. Harrison, C.E. Hubley, T.H.G. Jackson+, H.W. Johnston, W. Kay, W.C. Macready, R.S. Martin, J.A. McCallum*, N.H. McClelland, D.J. McDonald, D.S. McGeary+*, D.A. McKerricher, C.S. McLean, D.M. Mitchell+, T.J. Moe, L.B. Morrison, W. Moylan, R.V. Poole, J.M. Quigley, W.H. Simons+, D.C. Smith, J.W. Soloman, D. Storr+*, R.R. Sutherland, E.A. Thompson, F.A. Urquhart, J.K. Williams, M.A. Young.

No. 8 Course January 1943–April 1943
R. Bailey, J.G. Bell, H.G. Brewer, R.N. Chester, J. Clodman+*, D.W. Ellis, W.C. Feader, I.W. Findlay, W.K. Godfrey, D.N. Huntley, F.P. Ide, J.C. Jackson, A.A. Jeffrey, O. Johnson+*, W.W. Judd, C.B. McIntyre*, J.F. Quine*, S.J.W. Rickard, J.T. Rush, R.W. Swanick, O. Tropea, J. M. Wingfield*

No. 9A Course May 1943–August 1943
P.E. Ball (Miss), B.V. Benedictson*, G.H. Bevan, D.W. Boyd+*, R.H. Bridgman*, F.E. Burbidge+*, F. Carson+ (Miss), O.F. Carter, S. Christie, J.A. Doyle, A. Edwards, A.J. Filmer, J.E. Griffiths+ (Miss), W. Gutzman+*, D.H. Hanson, W.C. Hood, E.E. Lewis, W.W. Moorhouse+, A.H. Morrish, J.E. Parnall+, G.V. Patrick, K.G. Pettit*, B.A. Power+*, R. Rogers, I. Sheichet+, W.H. Smith+, H. Sussman, R. H. Swansburg*, J.H. Wells, E. Wybourn.

No. 9B Course May 1943–August 1943
H.T. Baker, E.R. Burford*, J. Carther, J.F.H. Connor, N.S. Dean, A.J. DesRoches, G. Donaldson, W. Doucet, J.R. Fowler*, C.A. Frank, E.S. Hallman, G. Higgins, H.W. How, G.H. Jeffrey, D.P. Kerr, L. Laking, W. Main, T.E.S. Mather, M. Mercer (Miss), J.A.E. Monette, J.M. Mulholland, M.A. Parker, B. Piercy, D.E. Rivers, P. Saltzman*, J.E. Samson, H.D. Skirrow, J.E. Tapsell, J.A. Townsend, F.H. Walden.

No. 10 Course November 1943–February 1944
C.C. Allen, R. Ayoub, D. Bishop, J.G. Bragg, H.J. Bucher, J. Crowley, A.R. Dahl*, R.B. Ferguson, B. Fleming, J.H. Greenhaugh, W.D. Grills, D.B. Henderson, N. Hunchak, C.F. Hunt*,

H.I. Hunter*, F.A. Innes, A.L. Jones, J.R. Mathieson+*, G.A. McKay+*, M. McKay (Miss), V. McWhinnie (Miss), R.A. O'Brien, J. O'Brien (Miss)*, R.H. O'Brien*, D.D. Oldrieve, C. Paulson, F.E.V. Paulson, M. Pirie (Miss), G.W. Robertson+*, R. Rosseland, W.D. Shales, J.M. Spiers, I. Smith (Miss), R.W. Smith, G. Stapley, C.H. Sutherland+*, H.A. Thompson+*, F.J. Turner, E.R. Walker+*, R.S. White, N. Wolverton, A.D. Wright.

No. 11 Course February 1944–May 1944
H.J.E. Dickin, J.W. Forrest, J.S. McGowan*, L.T. Millar*, W.D. Murden*, C.A. Mutton, R.S. Shannon*, O.C. Smith, L.E. Toombs, F.T. Upton*, C.K. Wilton, L.E. Woodbury.

No. 12 Course May 1944–August 1944
A.K. Brincomb, P.J. Denison+*, T.E. Hull+, R.C. Jacka, D.D. Levi, D.H. Parkinson*, F.L. Rogers, R.E. Sinclair, R.M. Skinner, C.C. Warkentin*, G.H. Washburn+*.

TABLE 2
Men who joined the Meteorological Division as observers and later were stationed at RCAF stations as metmen before the short courses were organized. Some subsequently took the advanced course while others continued as metmen throughout the war. (By year of entry.)

1934	D.W. Perrie
1937	D. Strachan, A.E. Allison
1938	W.G. Green, H. Cameron
1939	F.H. Patterson, H.P. Wilson, J. Labelle, N.N. Powe, C.E. Stevens
1940	P.J. Sandiford, W. Smith, L.J. Sobiski, V.G. Beirnes

APPENDIX 3

TABLE 3
Men who entered Meteorological Division as observers and later were stationed at RCAF stations as metmen for months before taking the short course. (By year of entry.)

1931	F.T. Upton
1937	D.S. McGeary, O. Johnson
1938	D.C. Day, A.R. Dahl, G.W. Robertson, H.A. Thompson
1939	R.S. Shannon, D.F.A. Slater, S.V.A. Gordon, C.B. McIntyre, J. Wingfield, E.A. Anderson
1940	N.V. Jefferson, D.E. Page, B.S.V. Cudbird, R.L. Titus, T.L. Richards, W.W. Allen, C.C. Warkentin, G.H. Washburn
1941	C.H. Sutherland
1942	J.R. Mathieson

TABLE 4
Graduates of the University of Toronto MA course in physics (meteorology) who were later employed by the Meteorological Division as meteorologists. (By year of graduation.)

1934	C.C. Boughner, F.G. Millar, M.N. Monsinger
1935	H.H. Bindon, E.W. Hewson, C.M. Mushkat, J.R.H. Noble
1936	D.H. Smith
1937	H.W. Edwards, D.B. Kennedy, F.J. Mahaffy
1938	A.M. Crocker, H.W. Halbert, R.A. Hornstein, D.E. McClellan, S.M. Neamtan, M.J. Oretzski, G.L. Pincock, D.M. Robertson, G.C.W. Tait, T.L. Wiacek
1939	R. Anderson, F.W. Benum, A.S.G. Grant, J.M. Leaver, D.P. McIntyre, K.T. McLeod, D.S. Ross, J.H. Sabraw
1940	S.J. Buckler, C.M. Cross, F.M. Kelly, R.W. Longley, L.B. MacHattie, J.A.D. MacNeil, H.L. Osmond, C.M. Penner, R.W. Rae, F.D. Thompson.

1941	H.D. Cameron, H.E. Chadburn, J.A. Conway, R.C. Graham, J.A. Lenahan, W.J. Noble, C.E. Thompson, J.A. Turner.

TABLE 5
Graduates of the University of Toronto and other universities who were hired as meteorologists by the Meteorological Division without taking the MA Course. (By year of entry.)

1925	E.B. Chilcott
1930	W.E.K. Middleton, A.R. McCauley, D.C. Archibald, H.P. Troop, A.F. Chisholm
1932	Andrew Thomson
1935	W.A. Thorn
1936	W.E.Turnbull, P.D. McTaggart-Cowan, R.C. Jacobsen, G.C. Gill, J.A. McPherson
1937	T.J.G. Henry
1938	T.G. How

APPENDIX 4

RCAF FLYING TRAINING SCHOOLS WITH METMEN, 1940–1945.

These dates may not agree with dates given elsewhere for the opening and closing of the flying training program at each station. The dates given here show the month when the first metman arrived at a station and the month when the last metman departed. Early in the war the flying training program began at some stations before a metman was available while later a metman was sometimes posted to a station before training began. Also, in 1944 and 1945, at the end of the training program, some metmen were kept at stations for weeks or months at the request of the RCAF to serve non-training units based at the station and itinerant pilots who used the airfield on occasion.

		From	To
Central Flying School			
	Trenton, ON	May 1940	continuing
Flying Instructors Schools			
No. 1	Trenton, ON	Aug 1942	Jan 1945
No. 2	Vulcan, AB	Aug 1942	May 1943
	Pearce, AB	May 1943	Jan 1945
No. 3	Arnprior, ON	Aug 1942	Aug 1942?
Service Flying Training Schools			
No. 1	Camp Borden, ON	Apr 1940	Jan 1946
No. 2	Ottawa, ON	Aug 1940	May 1946
No. 3	Calgary, AB	Oct 1940	Oct 1945
No. 4	Saskatoon, S K	Sep 1940	Jun 1945

No. 5	Brantford, ON	Nov 1940	Nov 1944
No. 6	Dunnville, ON	Nov 1940	Nov 1944
No. 7	Macleod, AB	Dec 1940	Nov 1944
No. 8	Moncton, NB	Dec 1940	Jan 1944
	Weyburn, SK	Jan 1944	Jun 1944
No. 9	Summerside, PE	Jan 1941	Jul 1942
	Centralia, ON	Jul 1942	Feb 1946
No. 10	Dauphin, MB	Mar 1941	Apr 1945
No. 11	Yorkton, SK	Apr 1941	Nov 1944
No. 12	Brandon, MB	May 1941	May 1945
No. 13	St. Hubert, QC	Sep 1941	Feb 1944
	North Battleford, SK	Feb 1944	May 1945
No. 14	Aylmer, ON	Jul 1941	Aug 1944
	Kingston, ON	Aug 1944	Oct 1945
No. 15	Claresholm, AB	Jul 1941	Mar 1945
No. 16	Hagersville, ON	Sep 1941	Sep 1945
No. 17	Souris, MB	Mar 1943	Mar 1945
No. 18	Gimli, MB	Sep 1943	Sep 1945
No. 19	Vulcan, AB	May 1943	Apr 1945
No. 31	Kingston, ON	Sep 1940	Mar 1941
		Apr 1943	Aug 1944
No. 32	Moose Jaw, AB	Aug 1941	Nov 1944
No. 33	Carberry, MB	Apr 1941	Nov 1944
No. 34	Medicine Hat, AB	Oct 1941	Nov 1944
No. 35	North Battleford, SK	Oct 1941	Feb 1944
No. 36	Penhold, AB	Jan 1942	Oct 1944
No. 37	Calgary (Currie), AB	Mar 1942	Mar 1944
No. 38	Estevan, SK	May 1942	Jun 1944
No. 39	Swift Current, SK	Dec 1941	Mar 1944
No. 41	Weyburn, SK	Feb 1942	Mar 1944

General Reconnaissance School

No. 1	Summerside, PE	Jul 1942	Sep 1945

APPENDIX 4

Operational Training Units

No. 1	Bagotville, QC	Jul 1942	Nov 1944
No. 3	Patricia Bay, BC	Nov 1943	Aug 1945
No. 5	Boundary Bay, BC	May 1944	Oct 1945
	Abbotsford, BC	Aug 1944	Oct 1945
No. 6	Comox, BC	Jun 1944	Jan 1946
No. 7	Debert, NS	Jul 1944	Dec 1945
No. 8	Greenwood, NS	Jul 1944	continuing
No. 31	Debert, NS	Jan 1942	Jul 1944
No. 32	Patricia Bay, BC	Oct 1941	Jun 1944
No. 34	Pennfield Ridge, NB	Jun 1942	Jul 1944
No. 36	Greenwood, NS	May 1942	Jul 1944

Wireless School

No. 1	Hamilton Mt. Hope, ON	Nov 1944	Jul 1945

Bombing and Gunnery Schools

No. 2	Mossbank, SK	Jul 1944	Dec 1944
No. 4	Fingal, ON	Jun 1944	Feb 1945
No. 7	Paulson, MB	Aug 1944	Aug 1945
No. 9	Mont Joli, QC	Dec 1941	Mar 1945

Air Navigation Schools

No. 1	Trenton, ON	May 1940	Oct 1940
	Rivers, MB	Nov 1940	Sep 1945
No. 2	Pennfield Ridge, NB	Jul 1941	May 1942
	Charlottetown, PE	Feb 1944	Jul 1945

Air Observer Schools

No. 1	Toronto Malton, ON	May 1940	Mar 1945
No. 2	Edmonton, AB	Aug 1940	Jul 1944
No. 3	Regina, SK	Oct 1940	Sep 1942
	Pearce, AB	Sep 1942	Apr 1943
No. 4	London, ON	Nov 1940	Dec 1944
No. 5	Winnipeg, MB	Feb 1941	Mar 1945
No. 6	Prince Albert, SK	Mar 1941	Sep 1942
No. 7	Portage la Prairie, MB	Aug 1941	Sep 1945
No. 8	Ancienne Lorette, QC	Sep 1941	Mar 1945
No. 9	St. Jean, QC	Jul 1941	Apr 1945
No. 10	Chatham, NB	Jun 1941	Mar 1945

Elementary Flying Training Schools

No. 3	St. Jean, QC	May 1945	Sep 1945
No. 6	Prince Albert, SK	Sep 1942	Nov 1944
No. 23	Yorkton, SK	Feb 1945	Oct 1945

Instrument Flying School

	Deseronto (Mohawk)	Apr 1943	continuing

INDEX OF NAMES OF PERSONS

INDEX

INDEX

INDEX

INDEX OF PLACES AND SUBJECTS

INDEX

INDEX

INDEX

INDEX

INDEX

INDEX

INDEX

INDEX